8 STRATEGIES FOR ENTERPRISE AI

AN EXECUTIVE'S GUIDE

With Real-World Case Studies

Alaa M. Tadmori

CONTENTS

INTRODUCTION

AI Without the Hype

The boardroom is tense. Your CFO slides a news article across the conference table and says, McKinsey says AI could add up to $4 trillion in annual economic value. What's our share going to be?[1] Your head of sales adds: Our biggest competitor just announced an AI initiative. Are we falling behind? Meanwhile, your CTO is bombarded with vendor calls promising revolutionary AI solutions that sound too good to be true.

The pressure to just "do something" is real. Boards demand "AI initiatives" without clear business cases, teams scramble to inject AI into any project superficially to show they're catching up, and organizations launch dozens of underfunded experiments that chase buzzwords instead of value. Wavestone's research reveals what executives won't say out loud: 75 percent of senior technology leaders admit to experiencing "generative AI FOMO," chasing competitors rather than strategy.[2] The result is pilot proliferation now filling a graveyard of half-baked experiments that have drained resources without delivering results.

So, to answer your question: "Am I already late to the AI game?" According to McKinsey's latest State of AI report, 55 percent of organizations have adopted AI in at least one function.[3] The race has definitely begun. But here's what matters: MIT's 2025 GenAI report found that 95 percent of generative AI pilots are failing to deliver business value.[4] Many don't even make it past proof of concept; Gartner predicts that at least 30 percent will be abandoned at that stage.[5] It's absolutely game on, but those who are pulling ahead are not necessarily those who started first. The window for strategic advantage is still wide open.

From my experience at Microsoft helping enterprises deploy AI at scale, I can tell you that most companies are still figuring it out ... and shedding millions of dollars along the way. AI without clear strategy is just expensive experimentation.

Most AI books give you one piece of the puzzle. Fix your data pipelines, restructure your teams, or secure leadership alignment. This book provides the full story: eight strategies that work as a holistic system. That's what separates the 5 percent of projects that succeed from the 95 percent that fail to deliver business value.

Before we get to strategy, a quick foundation.

The Shift That Changes Everything

The most profound change AI brings isn't technical, it's conceptual. Traditional IT systems are deterministic. Input A always produces output B. Your ERP system will calculate the same invoice total every time. This predictability is powerful, but limiting. Every possible scenario needs to be anticipated and programmed in advance.

AI systems are probabilistic. They recognize patterns and make predictions based on likelihood, not certainty. An AI reviewing expense reports doesn't follow a rigid decision tree; rather, it learns what legitimate expenses look like and flags anomalies with varying degrees of confidence. It might be "96 percent sure" that an expense is fraudulent, and "80 percent sure" another needs review.

This shift—from rules to patterns—fundamentally changes everything. With traditional systems, your primary concern was uptime: is it up and running? With AI, you must also monitor behavior. An AI system can be fully operational while quietly drifting toward biased outputs or confident hallucinations. Trace most AI failures and you will see them rooted in the classic up-and-running thinking.

This pattern-based approach is also why AI needs massive amounts of data to learn from. Think of it like teaching a child to identify birds. You don't give them a rulebook saying "has wings, feathers, beak." Instead, you show them hundreds of birds until they internalize what makes a bird a bird—even recognizing species they've never seen before. AI learns the same way, just with millions of examples instead of hundreds.

Take fraud detection. You train the model by feeding it thousands of historical transactions labeled as fraudulent or legitimate. The model learns what fraud looks like. Once trained, it's ready for inference— applying what it learned to new data. Show it a new transaction (could be one happening in real time), and it applies the patterns it learned to predict with a certain degree of accuracy, e.g., "a 92 percent probability of fraud." The model doesn't "remember" specific transactions from training; it has internalized the patterns that indicate fraudulent behavior.

This is why data quality is crucial. Feed an AI flawed data and it learns the wrong patterns. Feed it good data and it may uncover patterns no one thought were relevant. An AI is only as good as the data it lives on.

Your AI Instincts: The Questions That Matter

When someone pitched you a new IT initiative in the traditional IT world, two questions popped into your head automatically. "How will I measure success [i.e., the business impact]?" and "How long until this is live [i.e., the integration challenge]?"

These questions still matter. But in today's AI era, you need new instincts. When someone pitches a new AI initiative, develop the muscle memory to think of these additional questions:

- What data does this need, and do we have it clean and accessible?

- Can we actually act on what the AI predicts, or are we buying insights we can't operationalize?

- What happens when AI makes a wrong prediction? Can our business tolerate those errors?

Use such questions as a quick sanity check (though not a full evaluation framework, which I'll cover in Strategy 1). They will help surface obvious red flags before time is spent in deeper analysis. AI is just a tool, and knowing when not to use it is equally important as knowing when it's the right fit.

That is enough foundation. Now, strategy.

[1] "The Economic Potential of Generative AI: The Next Productivity Frontier."
McKinsey & Company. June 14, 2023.
https://www.mckinsey.com/capabilities/mckinsey-digital/our-insights/the-
economic-potential-of-generative-ai-the-next-productivity-frontier

[2] " Three-quarters of Tech Leaders Suffering from GenAI FOMO." Consultancy.uk.
April 9, 2024. https://www.consultancy.uk/news/36963/three-quarters-of-tech-
leaders-suffering-from-genai-fomo

[3] "The State of AI in 2023: Generative AI's Breakout Year." QuantumBlack: AI by
McKinsey. August 1, 2023.
https://www.mckinsey.com/capabilities/quantumblack/our-insights/the-state-of-
ai-in-2023-generative-ais-breakout-year

[4] Aditya Challapally, Chris Pease, Ramesh Raskar, and Pradyumna Chari. "The
GenAI Divide: State of AI in Business 2025." MIT NANDA. July 2025.

[5] "Gartner Predicts 30% of Generative AI Projects Will Be Abandoned After Proof of
Concept By End of 2025." Gartner [press release]. July 29, 2024.
https://www.gartner.com/en/newsroom/press-releases/2024-07-29-gartner-
predicts-30-percent-of-generative-ai-projects-will-be-abandoned-after-proof-of-
concept-by-end-of-2025

Think in Outcomes, Not Algorithms

Start with a Metric You Can Move

The AI could reveal the future. That's how it felt watching the demo. Fed with five years of sales data, it predicted next quarter's revenue down to the dollar. It spotted product trends before your merchandising team could. It even identified customers likely to churn, three months before they did so. The accuracy was spooky: 98 percent on every metric. Your team was mesmerized.

Six months and $12 million later, you're still waiting to see a dollar of business impact.

Sound familiar? Buried under all that technical sophistication is a simple truth: nobody has connected these impressive AI capabilities to actual business value. The vendor talks about what their AI can do, but not what problems it will solve. Your team is excited about the technology, not the outcomes it will deliver.

This scenario plays out across the Fortune 500 every day. I've watched brilliant executives, under pressure to "do something with AI," throw millions at solutions looking for problems. Companies get seduced by algorithmic elegance and miss the fundamental question that determines AI success: What value are we trying to create? What technology gets us there?

The companies winning with AI don't start with algorithms, they start with outcomes. They don't ask "What can this model do?" but "What business problem are we solving?" Success doesn't come from the biggest labs or flashiest demos. It comes from starting with a specific number you need to improve.

The Algorithm-First Trap

According to recent industry research, failure rates for AI projects remain stubbornly high. Gartner predicted that 30 percent of generative AI projects would fail to move beyond proof of concept.[1] A *Harvard Business Review* analysis found that some estimates place AI project failure rates as high as 80 percent—almost double the rate of corporate IT project failures from a decade ago.[2] Gartner identifies unclear business value as one reason GenAI projects are abandoned.[3]

These aren't technical failures. Most of these pilots work exactly as designed. They're strategic failures where technical success doesn't translate to business impact.

Here's how the algorithm-first trap typically unfolds: Your data science team discovers an interesting pattern in customer behavior data. They built a sophisticated model that predicts customer churn with 89 percent accuracy (an impressive technical achievement). They present it to you with enthusiasm: "We can predict which customers are likely to leave!"

Leadership approves the project, excited by the possibility of reducing churn. Congratulations, you have a beautiful dashboard showing churn predictions. The expensive reality check arrives later: customer retention rates haven't improved. The missing piece was obvious in hindsight: What specific actions will we take when we predict a customer is likely to churn? Do we have the operational capacity to execute those actions effectively?

That is the algorithm-first trap you need to spot immediately. Starting with what's technically possible instead of what's strategically necessary. It's the equivalent of buying a Ferrari to replace an old elevator. Impressive engineering, zero utility.

Consider the cautionary tale of IBM Watson for Oncology. IBM invested hundreds of millions to develop an AI system that could analyze cancer data and recommend treatments, marketing it as revolutionary cancer-fighting technology. They started with impressive AI capabilities rather than specific clinical outcomes. Internal documents later revealed that the company's medical specialists, along with its customers, had identified "multiple examples of unsafe and incorrect treatment recommendations."[4] Despite technical sophistication, Watson for Oncology was quietly discontinued by most hospitals because it solved theoretical problems rather than practical clinical challenges that oncologists actually faced.

IBM made a classic error: it fell in love with its algorithms' capabilities rather than its customers' problems. It built solutions in search of problems, rather than starting with problems and finding solutions.

According to McKinsey, the companies seeing the biggest bottom-line impact from AI—what McKinsey calls "AI high performers"—share key practices that others don't: they link their AI strategy directly to business outcomes rather than pursuing AI for its own sake.[5]

The Outcome-First Mindset

The outcome-first mindset is a deceptively simple strategy. Instead of asking "What can our data predict?" ask "What business decisions do we need to make better, faster, or more consistently?" Instead of showcasing technical capabilities, start with operational challenges.

I've advised countless CxOs who nod along when I share this. "Of course," they say, "business outcomes first." Then they go straight back to their teams to build impressive algorithms that solve nothing. I don't blame them; this strategy seems so obvious that they don't internalize how fundamental it is.

Let me share a historical parallel that helps make this memorable.

For over a thousand years, astronomers struggled with the impossibly complex mathematics of predicting planetary movements. In 929 CE, the brilliant astronomer Al-Battani made observations so precise they revealed a troubling truth. The math required to explain the planets' orbits was becoming absurdly complicated. Planets required "epicycles"—circles moving on circles moving on circles—to match observations.

Al-Battani spent his life trying to find more elegant mathematics, but he couldn't. The math remained torturously complex because he, like everyone else at that time, assumed Earth was the center of the universe.

Humanity wrestled with Al-Battani's complexity for six centuries until another genius—Nicolaus Copernicus, who studied Al-Battani's work obsessively—dared to ask: What if the math was complex because we put the wrong thing at the center? What if the sun, not Earth, was the anchor?

When he placed the sun at the center instead of Earth, the mathematical complexity disappeared. Planetary motion became elegant ellipses. Everything suddenly made sense.

AI strategy needs its own Copernican revolution.

When you put business outcomes at the center, the complexity vanishes. Technical metrics connect directly to business impact and value becomes self-evident.

Just as Copernicus didn't invent new math—he just changed what was at the center—you don't need more sophisticated AI to succeed at AI. What you need is to change what's at the center of AI strategy. You're not building AI solutions; you're solving business problems that happen to benefit from AI capabilities.

When adopting outcome-first thinking, you will become naturally better at spotting which AI opportunities matter and which are expensive distractions. You develop immunity to vendor pitches that lead with technical features instead of business value. Most importantly, you build organizational confidence in AI by delivering measurable results, not impressive zero-impact demos.

Don't forget that business outcomes should always include security and regulatory requirements. "Reduce processing time by 50 percent" isn't a complete outcome. "Reduce processing time by 50 percent while maintaining SOX compliance and established security requirements" is. Security and compliance objectives are often missing from online success stories, but they are an essential part of why the story was successful in the first place. These aren't constraints on outcomes; they're integral to them. An AI that achieves business goals but creates security vulnerabilities isn't a success. I'll come back to this in Strategy 7, for now remember that security and compliance are not something to be added after defining success. It's part of how success is defined.

Evaluating AI Opportunities

When navigating the constant stream of AI pitches and proposals, you need a comprehensive framework to separate genuine opportunities from expensive distractions. Here are four assessments that every AI opportunity must pass:

Business Value Assessment

Your outcome-first thinking becomes a filter. In the business value assessment, three questions cut straight to whether this AI serves strategy or will just showcase the tech:

1. **What specific actions will this AI perform in your day-to-day operations?**

   This forces specificity about action, not just information. If the answer is vague or focuses on "insights" rather than actions, you're looking at a solution searching for a problem. Here are some examples:

   > Vague answer: "It will give us insights into customer behavior."
   > Specific answer: "It will tell us which customers to offer retention discounts to, triggering our win-back campaign automatically when churn probability exceeds 70 percent."
   > Vague answer: "It will help us understand our supply chain better."
   > Specific answer: "It will decide which suppliers to reorder from based on demand forecasts, replacing our current manual weekly ordering process."

2. **What would happen if this AI solution didn't exist? What's the real cost of inaction?**

   This reveals whether you're alleviating real pain or inventing a problem. Quantify the pain of the status quo. If that can't be done, the problem isn't painful enough to justify the investment. Here are some examples:

Vague answer: "We'd miss out on AI transformation."
Specific answer: "We'd continue losing $2M monthly to fraud that human reviewers can't catch at current transaction volumes."
Vague answer: "Our competitors might get ahead."
Specific answer: "We'd keep eight analysts manually reviewing contracts for three days each, costing $400K annually in labor plus delayed deal closures."

3. **How will we know within 90 days whether this AI initiative is succeeding?**

This prevents endless pilots that never deliver value. You need specific metrics tied to business outcomes, not technical performance. And 90 days forces focus on value, not perfection. Here are some examples:

Vague answer: "We'll see if the predictions are accurate."
Specific answer: "Fraud losses will drop by at least 20 percent ($400K monthly) while false positive rates stay under 5 percent."
Vague answer: "Users will adopt the system."
Specific answer: "70 percent of customer service reps will resolve tickets 30 percent faster using AI suggestions, measured by our existing ticket system."

These three questions force clarity about value. Don't get trapped calculating exact ROI at this stage; that's premature. During evaluation, you need directional confidence, not decimal precision. But if you can't articulate value in rough terms—"this saves us approximately $2–3M annually" or "this prevents roughly 20 percent of quality defects"—then you're not ready to proceed.

Save the detailed ROI calculations for implementation, at which point you will have real data to work with. For now, focus on whether the opportunity has a credible path to value. Can you explain the financial logic to your CFO in two minutes? If not, don't proceed.

Only opportunities with strong answers to all three questions move forward to the next assessment. Why waste time on sophisticated AI that doesn't drive outcomes?

The Risk-Feasibility Assessment

Every AI opportunity falls somewhere on two dimensions—business value and implementation risk. High-value, low-risk opportunities get immediate attention. Everything else requires careful justification.

High business value means the AI addresses expensive problems or creates measurable revenue. Low implementation risk means having the data, skills, and operational capacity to deploy successfully.

Plot each opportunity on this 2x2 matrix:

- **High Value, Low Risk**: Proceed immediately.

- **High Value, High Risk**: Proceed with caution; worth pursuing but needs risk mitigation.

- **Low Value, Low Risk**: Consider for learning; good for building capabilities.

- **Low Value, High Risk**: Avoid; not worth the effort.

The Operational Readiness Assessment

Ask whether your organization can capture the business value this AI creates. The most sophisticated AI system is worthless if there is a lack of operational processes that can act on its outputs.

Example questions:

- If AI predicts customer churn, do you have retention programs ready to deploy?

- If AI identifies process inefficiencies, do you have the authority and resources to make changes?

- If AI spots quality issues, can your production line actually respond in real time?

- If AI recommends pricing changes, can your systems implement them quickly enough to matter?

The gap between AI capability and organizational readiness is where value goes to die. I will discuss how to build AI-ready organizations in Strategy 2.

The Governance Capability Assessment

Can you manage the risks this AI will create? A brilliant AI idea without proper governance is a lawsuit waiting to happen.

Key governance considerations:

- Can you explain the AI's decision-making to regulators?

- Do you have processes for when it fails?

- Are there clear lines of accountability when something goes wrong?

- Can you ensure fairness and prevent bias in its operations?

If the answer is "We'll figure out governance later," then pause. Many AI initiatives get cancelled at their first audit or regulatory inquiry.

Recognizing Where AI Solves Expensive Problems

Don't wait for vendors to pitch AI solutions. They don't know your business. The most successful AI implementations that I've seen came from executives who spotted expensive bottlenecks or patterns in their own operations, then evaluated whether AI was the right tool to solve them.

Here are common business problems that make good candidates for AI:

- **Look for volume with variance:** When you have thousands of similar cases that each require slightly different handling, you've found a potential match. Traditional automation fails here because it needs rigid rules. Human processing is expensive because of the volume. A major insurance company could cut claims processing from days to hours by using AI to handle thousands of claims daily because each one, though unique in its details, followed recognizable patterns. The AI could adapt to the variations that rigid automation can't handle.

- **Find expensive experts doing routine cognitive work**: In many investment banks, junior lawyers spend an enormous amount of time reviewing contracts for standard clauses. AI can handle the first-pass review—flagging unusual terms and summarizing key provisions—while lawyers focus on negotiation and strategy. The same pattern appears in many industries: radiologists screening routine mammograms, consultants creating industry benchmark reports, auditors checking expense reports against policy.

- **Target problems that produce clear, measurable pain**: The best problems to solve have unambiguous costs. Processing delays that can be timed. Error rates that can be counted. Costs per transaction that can be calculated.

- **Start where expensive mistakes won't kill you**: Look for problems where errors are costly but recoverable. These create natural learning environments where AI can improve without catastrophic risk. Major retailers use AI for grocery substitution recommendations, like suggesting another type of soap when the requested brand is out of stock. Incorrect substitutions may annoy customers and trigger refunds, but each mistake teaches the system. The cost is clear but not catastrophic.

Case Study: BMW's Million-Dollar Stud Problem

BMW's Spartanburg, South Carolina, plant cranks out 1,500 vehicles daily.[6] In the body shop alone, robots weld up to 400 metal studs onto every single SUV frame (half a million studs daily).[6] When even one stud lands in the wrong spot, production grinds to a halt. A technician has to hustle over and manually fix the error while million-dollar machinery sits idle.

If BMW had fallen into the algorithm-first trap, here's what would have happened. Some vendor would have strolled in with a dazzling demo. "Let us revolutionize your factory with our Industry 4.0 platform!" They'd showcase AI systems with 99.7 percent accuracy rates, promise fully autonomous production lines, paint visions of lights-out manufacturing where robots run the show 24/7. The price tag would have been only $100 million and three years to completely rebuild production systems.

BMW asked one simple question: "What specific problem costs us the most money right now?"

The answer wasn't glamorous. It was studs landing in the wrong place. That's it.

So, BMW added a single AI-powered laser system that scans for misplaced studs, then tells the existing robots—the same robots BMW has been using for years—how to fix the mistake. No massive overhaul. Just one smart addition that eliminated a bottleneck.

That laser system saves BMW over $1 million annually.[6] And they're now achieving five times the efficiency they thought was possible.[6] Not because they revolutionized manufacturing, but because they removed the friction that was holding back their existing excellence.

Had BMW pursued algorithm-first thinking, it would probably still be in year two of a "comprehensive smart factory transformation".

The takeaway is simple: Stop asking what AI can do. Start by asking "What expensive problem keeps hitting our margins?"

STRATEGY 1 ACTION PLAN

Days 1-30: Monday morning, every AI initiative gets a business value reality check. Pull up every AI initiative in your organization and run each one through the four AI assessments. Review actual results, concrete numbers, not projections. You'll likely discover a mix of initiatives with clear business value and others that are really just teams learning new technologies. Don't automatically kill the experiments—learning has value—but fund them appropriately and set different expectations.

After auditing, model a simple change: when anyone updates you on AI initiatives, first ask about business impact.

Days 31-60: Walk through operations, physically if possible, mentally if not. Hunt for volume-with-variance problems, expensive experts stuck doing pattern matching, processes with clear measurable pain, and areas where mistakes are costly but recoverable. Pick a specific bottleneck that makes your CFO gasp. That's your next AI opportunity. Run it through the four assessments before committing resources.

Make this non-negotiable: every new AI initiative must pass the four assessments before getting a dollar of funding. Train teams to apply this framework in their evaluation process.

Days 61-90: Watch for the warning signs: technical metrics creeping into business reviews, pilots multiplying while production deployments are rare, teams pitching AI capabilities instead of business solutions.

When you spot drift, redirect conversations back to business value. The expectation should be clear: AI discussions lead with money saved or revenue generated.

[1] "Gartner Predicts 30% of Generative AI Projects Will Be Abandoned After Proof of Concept By End of 2025." Gartner. Press Release. July 29, 2024. https://www.gartner.com/en/newsroom/press-releases/2024-07-29-gartner-predicts-30-percent-of-generative-ai-projects-will-be-abandoned-after-proof-of-concept-by-end-of-2025

[2] Iavor Bojinov. "Keep Your AI Projects on Track." *Harvard Business Review*. November-December 2023. https://hbr.org/2023/11/keep-your-ai-projects-on-track

[3] "The Economic Potential of Generative AI: The Next Productivity Frontier." McKinsey & Company. June 14, 2023. https://www.mckinsey.com/capabilities/mckinsey-digital/our-insights/the-economic-potential-of-generative-ai-the-next-productivity-frontier

[4] Casey Ross and Ike Swetlitz. "IBM's Watson Supercomputer Recommended 'Unsafe and Incorrect' Cancer Treatments, Internal Documents Show." *STAT+*. July 25, 2018. https://www.statnews.com/2018/07/25/ibm-watson-recommended-unsafe-incorrect-treatments/

[5] "The State of AI in 2022—and a Half Decade in Review." McKinsey & Company. December 6, 2022. https://www.mckinsey.com/capabilities/quantumblack/our-insights/the-state-of-ai-in-2022-and-a-half-decade-in-review

[6] Jake Callahan and Andrea Day. "How BMW Uses A.I. to Make Vehicle Assembly More Efficient." CNBC. July 21, 2023. https://www.cnbc.com/2023/07/21/how-bmw-uses-ai-to-make-vehicle-assembly-more-efficient.html

Close the Gaps That Stall AI
Tech Isn't the Problem. Your Org Is

Clear business outcomes had been set, success metrics established, and the right team hired. The built tech was impressive. It returned 94 percent accuracy on demand forecasting, which tied into inventory cost reduction targets. Yet, the AI initiative failed to deliver ROI.

The problem was something no vendor mentioned, nor any keynote speaker addressed. The organization simply wasn't ready for AI.

Data was locked in departmental silos, guarded by VPs who treated information like personal currency. The culture punished failure, so teams didn't even bother testing the solutions. The infrastructure creaked under basic workloads, let alone real-time AI processing. And while the executive team gave rousing speeches about 'becoming an AI-first company,' none of them had attended a single AI steering committee meeting.

This scenario plays out across organizations worldwide. Research consistently shows that organizational factors, not technical limitations, are the primary barriers to AI success.[1] The truth is that AI will only be as good as an organization's ability to use it. The smartest data scientists can be hired, the most sophisticated platforms purchased, and partnerships established with leading AI vendors, but if an organization isn't ready—culturally, operationally, structurally—cash will go up in smoke on proof-of-concepts that never scale.

The good news is that organizational readiness is entirely within your control. No need to wait for the next AI breakthrough to start putting it into practice. This chapter covers how to assess readiness, identify operational gaps, and build an organizational foundation for success. If not in a position to transform your organization, this framework can protect you. When asked to lead an AI initiative, run this assessment first. If multiple dimensions score below three on the worksheets that follow, there are two choices: secure written commitments to fix them, or politely decline. Brilliance and technical expertise won't overcome organizational dysfunction. Protect your reputation. Rather than become another failure, explaining to the board why $12 million disappeared with nothing to show for it, just pass on a project that is doomed to fail.

The Organizational Readiness Framework

Systematic assessment of organizational readiness is required before spending another dollar on AI technology. Think of it like preparing for a marathon. You wouldn't start by buying the most expensive running shoes and immediately attempt 26.2 miles. Instead, the current fitness level would be assessed, training realities noted, and capacity would then be built systematically.

The five dimension-readiness frameworks I'm about to share are based on established organizational change and digital transformation principles. Each is necessary but not sufficient on its own. And the impact of these dimensions doesn't add, but multiply. Top talent with misaligned leadership will fail. Great infrastructure with poor data will fail. Strong leadership with weak culture will fail. Excel in four, but fail in one and that single weakness will cripple the entire AI program.

Here's the assessment scale for each dimension:

- **Level 1. Unprepared**: Significant barriers exist that will block AI success.

- **Level 2. Aware**: Problems are recognized but not yet addressed.

- **Level 3. Developing**: Active efforts underway, but not yet mature.

- **Level 4. Capable**: Solid foundation with some room for improvement.

- **Level 5. Advanced**: Organization excels in this dimension.

The goal isn't perfection; that's unrealistic and unnecessary. The goal is understanding where you stand, identifying what gaps will block AI ambitions, and understanding what to fix first.

First Dimension: Leadership Alignment

There's a difference between declaring AI a priority and treating it like one. Executives who champion AI in quarterly meetings but never follow up, no questions about progress, no review of metrics, no defense of the budget when cuts arrive aren't leading. They're performing. Performative support evaporates when real commitment is required. Research from the *MIT Sloan Management Review* consistently shows that successful digital transformation requires leaders who combine clear vision with active engagement. As MIT Sloan's George Westerman puts it, "The ability to envision and drive change is just as important as the ability to work with technology. If you don't have both, you can't succeed in this world."[2]

Effective AI leadership requires different behaviors than traditional executive sponsorship. Leaders need to not just talk about AI, but fundamentally restructure how they engage, and require senior executives to actively participate in AI transformation efforts. AI leadership isn't just about the CEO, but alignment throughout the leadership chain. Even when senior leadership is supportive, resistance from middle management can kill AI initiatives. Managers can see AI as threatening their authority, jobs, or carefully constructed departmental empires.

Leadership alignment also requires understanding security and governance implications. Every readiness dimension must incorporate responsible AI principles from the onset. Leaders who champion AI without considering security and governance create risk, not readiness.

Leadership Alignment Assessment

Answer these questions (one point for each "yes"):

- Can each leader articulate specific AI impacts on their division over the next 24 months?

- How many hours, per month, does each leader spend directly engaged with AI initiatives?

- Do leaders personally engage when AI hits obstacles, or do they delegate to someone else or some committee?

- Are leaders' performance metrics tied to business outcomes from AI initiatives?

- Are AI initiatives protected as strategic investments during budget pressures?

A score below three highlights leadership alignment gaps that will derail AI efforts. Unlike technical or data challenges, leadership misalignment can move invisibly making it harder to fix.

Building Genuine Leadership Alignment

Start with honest assessments. At the next leadership meeting, ask, "If our AI initiatives succeed wildly, what happens to your role in three years?" The responses will reveal true readiness. Leaders prepared for AI give answers like "My team will shift from processing to exception handling" or "My role will evolve from managing processes to managing outcomes." Those who aren't ready will focus on why AI can't replace human judgment in their domain.

Real leadership alignment requires four commitments.

1. **Time investment:** Every AI initiative, including AI pilots, needs a clear owner from day one. Not a committee, not "shared accountability," but one person who owns the results, whether that's a win at the board meeting or a crisis on a Sunday morning. AI needs sustained attention, not check-the-box reviews. Leaders must engage regularly with AI teams to understand challenges and remove obstacles. But engaging isn't enough if leaders lack the AI fluency to contribute meaningfully. AI literacy is vital.

2. **Functional AI literacy:** Leaders who champion AI without understanding its basics approve (or disapprove) projects without evidence and can't distinguish vendor hype from genuine capability. Companies excelling at AI adoption have broadly distributed AI literacy across the leadership team, not concentrated in the CTO or CIO. This doesn't mean leaders should learn data science. It means enough functional knowledge to ask the right questions.

3. **Risk tolerance:** AI involves experimentation and failure. Leaders must model learning from failures, not punishing them. This also means accepting that a portfolio approach works better than betting everything on one initiative. Organizations that concentrate all AI investment in a single high-profile project are one failure away from leadership declaring the whole effort a

waste. Risk should be spread across a few targeted initiatives so that learning from one feeds into the others.

4. **Sustained investment**: Protecting the AI budget when critics question spending is essential. Funding must also be fixed and transparent from the start. Organizations that fund AI with enthusiasm during the honeymoon period then quietly reallocate resources six months later are setting up their teams for failure.

Second Dimension: Cultural Readiness

The data science team presents their breakthrough to regional sales directors. "We can reduce customer churn by 25 percent, worth $3 million annually. We just need access to your regional sales data to train the model for each market."

The room temperature drops. The first objection comes quickly.

"My team spent years building those customer relationships. An algorithm can't replicate what they know."

Another leader nods. "Churn in my region is driven by factors no dataset captures—competitor relationships, local market dynamics. A model won't see that."

A third delivers the killing blow. "That data contains sensitive customer information. Legal would never approve sharing it."

In thirty seconds, cultural resistance killed an initiative that could have saved millions. The business case was clear. But the culture wasn't prepared for AI's collaborative, experimental nature.

Traditional organizational culture rewards certainty, punishes failure, protects information, and preserves hierarchies. It evolved to excel at the opposite of what AI requires. AI thrives on experimentation, learns from failure, requires data sharing, and flattens decision-making.

The Cultural Readiness Assessment

Answer these questions (one point for each "yes"):

- **Experimentation comfort**: Can teams try new approaches with minimal or no approvals?

- **Data accessibility**: Can analysts access cross-functional data instantly or within hours, not weeks?

- **Failure tolerance**: Is learning from failure rewarded or punished?

- **Collaboration patterns**: Do teams naturally work across functional boundaries?

- **Change adaptability**: How quickly can the organization adopt new tools and processes?

If your culture scores below three, the following interventions are starting points:

1. **Innovation sandboxes**: Create bounded spaces where different rules apply, experimentation is expected, failure is learning, and data flows freely.

2. **Learning from failure sessions**: Regularly showcase failed experiments. This transforms failure from something hidden to something that advances organizational knowledge.

Third Dimension: Data Foundation

Having data isn't the same as having AI-ready data. Most enterprise data was accumulated without AI in mind, creating three critical gaps:

1. **Accessibility paradox**: Data exists, but teams need six approvals and three weeks to access it.

2. **Quality issues**: Data exists, but it's not clean enough for AI. Inconsistent formats, missing fields, duplicate records. According to DAMA-DMBOK, data quality has six dimensions: accuracy, completeness, consistency, timeliness, validity, and uniqueness. AI amplifies every flaw.

3. **Integration challenges**: Data exists, but scattered across systems with different definitions and schemas. Master data management and semantic consistency matter more than physical consolidation.

Data Readiness Assessment

For each of the five critical entities—customer, product, employee, supplier, transaction—add a point for each (for a total of 25 possible points) if data is:

- Consolidated in one system (not scattered).

- Consistent across systems.

- Updated at least daily.

- Accessible within 24 hours.

- Well-integrated with other systems.

A score below 15 signifies existing data barriers that will sabotage any AI initiative. Strategy 3 provides the complete framework for building a strong data foundation but for now, understand that poor data quality doesn't just reduce AI accuracy. Your best people end up spending 80 percent of their time making sense of data, rather than creating value. Even worse, poor data quality silently erodes trust. When AI confidently recommends keeping an important customer on hold while prioritizing a fraudster; teams will stop believing in the technology altogether.

Fourth Dimension: Talent and Capability

The talent challenge runs deeper than hiring data scientists. According to the World Economic Forum's *Future of Jobs Report*, 63 percent of employers identify skill gaps as a major barrier to AI adoption—and the gap isn't just technical AI skills. The demand for professionals who can bridge technological capabilities with business value is growing. Combining AI literacy with creative thinking, resilience, and adaptability is the new standard.[3]

Successful AI requires AI translators, people who speak both business and technical languages fluently. Without this bridge, technical teams build the wrong thing and business teams can't evaluate what they're getting.

Talent Readiness Assessment

Answer these questions (one point for each "yes"):

- **Business translation**: Can technical and business teams communicate effectively?

- **Technical depth**: Is there sufficient specialized expertise for your ambitions?

- **Implementation capability**: Do you have people who can operationalize AI outputs?

- **Learning infrastructure**: How quickly can new AI capabilities be acquired?

If the score is below three, talent gaps will bottleneck potential AI solutions. Strategy 4 will provide a complete playbook for building A-teams that actually deliver but for now, know the talent problem isn't about finding unicorn data scientists. It's about empowering the analytical minds already in the organization, the people who understand the business but need the tools and structure to drive AI value.

Fifth Dimension: Technical Foundation

AI infrastructure isn't just compute. It's how data flows, how systems connect, and how new capabilities integrate with what you already have. Assess what you have before assuming you need something new.

Technical Readiness Assessment

Answer these questions (one point for each "yes"):

- **Data accessibility**: Can AI models access needed data?

- **Processing capability**: Are there sufficient computing resources for the use cases?

- **Integration flexibility**: Can new systems work with existing infrastructure?

- **Security and compliance**: Can AI workloads run without violating regulatory or security requirements?

- **Operational reliability**: Can the infrastructure handle AI workloads without degrading the performance of the existing system?

A score below three highlights technical gaps that will limit what is possible.

Here are three pragmatic patterns that will expand what is possible:

1. **API enablement**: Create modern APIs that expose legacy data safely.

2. **Edge computing**: Deploy AI models locally when cloud solutions aren't feasible. Manufacturing companies run quality control AI on factory servers, and then only send the results to the cloud.

3. **Hybrid architecture**: Gradual migration with value proofing at each step. Move specific functions to modern infrastructure while maintaining critical operations on legacy systems.

The Big Picture

With all five dimensions assessed, a complete picture of an organization's readiness for AI is now available.

Notice that this framework applies to any major digital transformation, including cloud migrations, or company-wide ERP systems. In fact, the most successful AI organizations are those with mature digital transformation practices already in place.

With AI, these same principles are amplified. The complexity is greater and the margin for error smaller. If an organization struggled with a cloud migration or ERP integration, AI will expose the same weaknesses that surfaced with them, only faster and more painfully. But if strengthened processes were already built during previous transformations, you're halfway to AI readiness.

Case Study: Domino's Pizza's Readiness

In 2010, Domino's faced a brutal reality. Customer feedback was brutal: "Crust tasted like cardboard, sauce like ketchup, totally void of flavor".[4] Sales had plummeted, with domestic same-store sales falling 10 percent between 2006 and 2008. Patrick Doyle, promoted to CEO in March 2010, had a choice.[5] He could follow the traditional playbook of hiring consultants, rolling out a multi-year turnaround program, and decommissioning legacy systems. Instead, he did something that would make most executives break out in hives. He went on national television and admitted their pizza sucked.

"You can either use negative comments to get you down, or you can use them to excite you and energize your process to make a better pizza," Doyle said in the now-famous video. "We did the latter."[6]

But bad pizza doesn't fix itself. A failing product is usually a symptom of a failing organization. New recipes mean nothing if the people, systems, and processes can't execute them consistently across thousands of locations. Domino's didn't just fix the pizza. They rebuilt the organization behind it.

Doyle started this with leadership alignment. He didn't delegate digital to a Chief Digital Officer in some satellite office. He made technology central to everything. When smartphones started replacing flip phones, Doyle challenged his IT team with a specific goal: enable customers to complete a pizza order while waiting at a stoplight.[9] Think about that precision. Not "improve mobile ordering." Not "enhance the customer experience." The 30 seconds it takes a stoplight to change.

By 2015, Doyle was telling anyone who'd listen.

"We believe by transaction counts we're in the top five of e-commerce companies in the world. We know we're behind Apple and Amazon, but we may be as high as third," explained Doyle to Marketplace.[7] When the CEO is comparing you to Amazon, not Pizza Hut, that changes everything.

The cultural transformation was even more radical. Domino's headquarters in Ann Arbor employed about 700 people. By 2015, nearly half of them, about 300 employees, worked in technology.[8] Not marketing or operations, technology. This wasn't bolting digital onto a pizza company. This was becoming a technology-first company.

Changing headcount doesn't change culture. So, Doyle instituted something brilliant. Every employee above store level—accountants, marketers, lawyers—had to train in a functioning Domino's store built inside its headquarters. "You're going to learn the basics," Doyle explained, "and you're going to learn about how what you do day to day affects the stores."[7]

The culture shift went deeper. Kevin Vasconi, who joined as CIO from Stanley Black & Decker, said, "What a great opportunity as a technologist to work for a company where the CEO is already convinced that technology can change the business." Four years earlier, Vasconi struggled to recruit talent. By 2016, he noted, "I used to spend a far larger amount of time just selling people on the company." Now? "If you're an IT professional, you're in a candy store."[9]

For data foundations, Domino's made a critical decision early. One point-of-sale system for every store.[6] Some franchisees sued, wanting their own systems, but Domino's held firm. This unified infrastructure meant every pizza order, every customer interaction, every delivery time fed into the same system. When launching Pizza Tracker, which lets customers follow their order from placement to delivery, it wasn't a bolt-on feature. It pulled data directly from the POS system that was already tracking everything.

The talent transformation was remarkable. Domino's didn't just hire data scientists and hope for the best. They built cross-functional teams where IT worked directly with operations. They created a two-year technology rotation program for new graduates, rotating them through four different areas of the business.[10] When technology talent is treated as future business leaders, and not just coders, different results follow.

On the infrastructure side, Domino's approached it very pragmatically. They didn't rip out everything and start over. They kept their delivery system but they built a tech layer on it that made ordering frictionless. The "Easy Order" feature that saved customer preferences became the foundation for ordering via emoji, smart TVs, Ford cars, and eventually voice assistants.[8]

The stock price was below $10 when Doyle started, it exceeded $200 by the time he left in 2018.[6] By 2018, Domino's had become the largest pizza company in the world, overtaking Pizza Hut.[11]

The lesson isn't that every company should admit their product sucks on national television. The lesson is that transformation requires aligning every dimension of organizational readiness with the unified goal.

Compare this to companies that fail at digital transformation. They hire a Chief Digital Officer (leadership theater in most cases). They talk about "becoming more agile" (buzzwords). They launch data lakes that never deliver value (large scale without purpose). They recruit PhDs who leave after six months (talent without integration). They buy the latest platforms (solutions without problems). But, organizational readiness isn't about having the most advanced technology or the biggest transformation budget. It's about aligning the entire organization—leadership, culture, data, talent, and infrastructure—around a clear vision and executing relentlessly. When that is done right, you can transform cardboard pizza into a tech powerhouse.

Days 1-30: This week, run all five assessments. Document scores; no interpretation. Identify your weakest dimension—that's your first priority.

Assign a single owner to each existing AI initiative. Not a committee, one person.

To build functional AI literacy across the leadership team, establish a monthly working session where leaders work directly with AI practitioners on real initiatives, reviewing actual model outputs and challenging assumptions, not watching presentations about AI.

Days 31-60: Fix gaps based on scores. Leadership below three: get each leader to commit specific hours and resources to AI; lock in AI budgets with explicit written protection from mid-year reallocation. Culture below three: launch an Innovation Sandbox and schedule a learning-from-failure session. Data below 15: fix customer data consistency. Infrastructure below three: create APIs to expose critical legacy data.

Days 61-90: Train five business professionals on AI basics (not data science). Run a small, low-risk AI pilot to track organizational metrics: data access time, leader engagement, and failure tolerance. If the pilot showed organizational readiness, scale. If not, fix gaps first.

If running only one AI initiative, identify a second low-risk candidate. A single initiative is a single point of failure for organizational confidence.

[1] Jin Li, Feng Zhu, and Pascal Hua. "Overcoming the Organizational Barriers to AI Adoption." *Harvard Business Review*. November 2025. https://hbr.org/2025/11/overcoming-the-organizational-barriers-to-ai-adoption/

[2] Tam Harbert. "Digital Transformation Has Evolved. Here's What's New." MIT Sloan School of Management. May 18, 2021. https://mitsloan.mit.edu/ideas-made-to-matter/digital-transformation-has-evolved-heres-whats-new

[3] "The Future of Jobs Report 2025." World Economic Forum. January 2025. https://www.weforum.org/publications/the-future-of-jobs-report-2025/digest/

[4] Domino's Pizza. "Domino's Pizza Turnaround." YouTube video. 2009. https://www.youtube.com/watch?v=AH5R56jILag

[5] Bret Thorn. "Patrick Doyle Leaves Legacy of Success at Domino's." *Nation's Restaurant News*. January 11, 2018. https://www.nrn.com/restaurant-executives/patrick-doyle-leaves-legacy-of-success-at-domino-s

[6] Jonathan Maze. "How Patrick Doyle Changed Domino's, and the Restaurant Industry." *Restaurant Business*. June 25, 2018. https://www.restaurantbusinessonline.com/leadership/how-patrick-doyle-changed-dominos-restaurant-industry

[7] Marketplace. "Domino's CEO Patrick Doyle: Tech with a Side of Pizza." September 24, 2015. https://www.marketplace.org/2015/09/24/business/corner-office/dominos-ceo-patrick-doyle-tech-side-pizza/

[8] Ron Ruggless. "The Power List 2016: No. 6 Kevin Vasconi." *Nation's Restaurant News*. January 19, 2016. https://www.nrn.com/power-list-2016-no-6-Kevin-Vasconi

[9] Jonathan Maze. "How Domino's Became a Tech Company." *Nation's Restaurant News*. March 28, 2016. https://www.nrn.com/restaurant-technology/how-domino-s-became-a-tech-company

[10] Domino's Careers. "Development Programs." Accessed January 24, 2026. https://jobs.dominos.com/us/explore-dominos/development-programs/

[11] Chris Matyszczyk. "Domino's Just Overtook Pizza Hut as the World's Biggest: Here's the 1 Reason Pizza Hut Lost." *Inc.* February 21, 2018. https://www.inc.com/chris-matyszczyk/dominos-just-overtook-pizza-hut-as-worlds-biggest-heres-1-reason-pizza-hut-lost.html

Make Data a Strategic Asset
| Business Sets the Rules, Not IT

It was the 2013 holiday shopping season, the most critical week for retail in the year. For three weeks, hackers extracted 40 million credit card numbers and 70 million customer records from one of America's most trusted retailers. The root cause was a third-party HVAC vendor's credentials that provided the initial access, but the real failure was deeper. Target's data architecture allowed attackers to move laterally from air conditioning systems to payment processing to customer databases. Perfect algorithms couldn't overcome the fundamental flaw of a data foundation built for efficiency, not resilience.

Target had invested heavily in advanced analytics. Famously, it could predict when customers were pregnant based on purchasing patterns, with algorithms sophisticated enough to sometimes send targeted coupons that revealed pregnancies before families even knew. Yet all that algorithmic sophistication meant nothing when its data foundation crumbled.

The breach cost Target's CEO and CIO their jobs. It also cost the company $162 million in breach-related expenses, and it later paid additional settlements including $18.5 million to states, $10 million to resolve a federal class action, $39 million to U.S. banks, and $67 million to Visa.[1] Worse yet, it lost trust. Customers fled to competitors, which led to a 46 percent drop in holiday season profits.[2] A stark lesson. In the AI age, data foundation isn't just about analytics. It's what makes or breaks you.

A note before diving in. This chapter (and the Strategy 7 chapter) is more technical than others, and deliberately so. Executives who don't understand the fundamental technical principles can't ask the right questions. The data team will handle implementation specifics, but you need to understand what a modern architecture looks like and why it matters for AI decision-making.

Security: The Overlooked Profit Center

Target's $162 million example teaches a precious lesson. Data is the foundation that determines whether you build tomorrow's competitive advantage or tomorrow's headline-grabbing breach.

The good news is that most breaches aren't sophisticated attacks, but predictable failures of basic data hygiene.

Build Security Within, Not Upon

Building security within means making it part of every architectural choice. Security built into data architecture does cost more upfront; however, it saves exponentially by avoiding the costs of breaches, audits, and retrofitting. When security is an afterthought, it delays every project. When it is woven into the platform and operations, every project becomes secure by inheritance.

Every data decision must pass four critical tests:

- **Encryption**: Protects data both when stored and in transit. Encryption is non-negotiable.

- **Access control**: Based on least privilege, meaning each employee gets the minimum access required for their role, not what's convenient for IT to manage. Marketing can see customer preferences but not credit card numbers. Sales can view purchase history but not social security numbers.

- **Containment strategy**: Limits damage when breaches occur. Target's disaster wasn't just the initial HVAC breach; it was that attackers were able to move laterally from air conditioning to payment processing to customer databases. Architecture must be designed with security boundaries, like fire doors in a building, meaning network segmentation between systems, separate credentials for different environments, and ensuring a breach in marketing analytics can't reach payment processing.

- **Audit trails**: Prove governance works. Every data access, every permission change, every unusual query pattern must be logged and monitored. When regulators ask who accessed customer data last Tuesday at 1:00 PM, the answers are needed in minutes, not days.

A good data foundation is a core requirement for companies to succeed in AI. A *Harvard Business Review* study found that top-performing data-driven companies outperformed competitors by 5 percent in productivity and 6 percent in profitability.[3] This performance difference remained robust even after accounting for labor, capital, purchased services, and traditional IT investment. The message is clear. Data-driven companies win.

Yet, despite this evidence, organizations struggle to capitalize. The 2024 Wavestone study found that less than half—48.1 percent—of Fortune 1000 organizations have successfully built a data-driven organization.[4] Research by McKinsey Global Institute reveals that data-driven organizations show dramatic advantages, yet the same research shows that most companies capture only a fraction of data's potential value.[5]

Considering that AI lives on data, this is a problem. Months spent wrestling with data quality are months during which AI initiatives remain stalled.

The Economics of Data

Most organizations still treat data as exhaust—a byproduct of operations. They invest in data infrastructure, not data quality. The cost of that gap is staggering.

According to research from Coherent Solutions, 15–25 percent of total AI project costs go to data acquisition and preparation.[6] On a $10 million AI initiative, that's $1.5–$2.5 million just to get data ready, never mind talent, innovation, or basic preparation.

IBM estimated that poor data quality costs the US economy a staggering $3.1 trillion annually.[7] According to Gartner, poor data quality costs organizations an average of $12.9 million per year.[8] The impact, of course, varies according to an organization's size and data complexity. For a $1 billion company, even a conservative 1–2 percent revenue impact from data quality issues translates to $10-20 million annually. Factor in hidden costs like wasted time, missed opportunities, compliance failures, and the real number climbs higher.

This aligns with research published in the *Harvard Business Review*, which found that "knowledge workers waste 50 percent of their time hunting for data, identifying and correcting errors, and searching for confirmatory sources for data they don't trust"[9] — a reflection of systemic data failures, not employee inefficiency. The hidden cost is that decision-makers at all levels routinely act on bad data without even knowing it. Consider the cascade effect of poor data quality:

- **Direct costs**: Include time spent finding, cleaning, and reconciling data.

- **Opportunity costs**: Encompass delayed decisions and missed market windows.

- **Risk costs**: Stem from bad decisions based on bad data.

- **Reputation costs**: Arise from customer trust lost due to data errors.

- **Regulatory costs**: Come from fines and penalties for data-related compliance failures.

Every manual reconciliation is money burned monthly. Every time five teams calculate customer lifetime value differently, you're paying for confusion. Track these hidden costs for one week; the number will be a shock.

The Four-Stack Trap: Why Enterprise AI Takes Forever

Poor data quality is a symptom of a deeper architectural problem that makes enterprise AI complex, costly, and slow. Ask a CTO how many different technology stacks process data. They'll describe at least four incompatible systems that barely communicate, each built for a different era, each speaking a different language, each requiring specialized skills.

Here's what accumulated in most enterprises over the past 30 years:

- **Analytics stack**: Handles "What happened?" scenarios. Built in the 1990s around data warehouses like Teradata or Oracle, these systems excel at structured reporting. They process data in nightly batches, generate quarterly reports, and feed executive dashboards. But trying to get real-time insights or handle unstructured data was impractical. These systems were built when "fast" meant overnight processing and "big data" meant gigabytes, not petabytes.

- **Streaming stack**: Built to answer "What's happening now?" Added in the 2010s, when real-time became critical, platforms like Kafka and Kinesis process millions of events per second. Website clicks, IoT sensors, and transaction streams flow here. But this stack can't easily access historical patterns. It knows what happened one second ago but not last year. The data formats are different, the tools are different, and the skills are different.

- **Engineering stack**: Makes sense of chaos. ETL pipelines, business rules, and data transformations live here in a sprawl of custom code, stored procedures, and integration platforms.

- **ML stack**: Predicts "What will happen?" next. The newest systemic additions, platforms like Azure ML or SageMaker, require data in specific formats, such as tensors, vectors, and feature stores. Data scientists build models here, but getting these into production means reverse-engineering connections to the other three stacks, like building a Rolls engine but having to attach it to a horse carriage.

No one intended to build four stacks. Everyone was just solving business problems as they arose, siloed in their own space. Each decision made sense at the time. Together, they created an architectural nightmare. The real cost becomes clear when attempting enterprise AI.

Take a bank trying to detect fraud in real-time as an example. A straightforward use case that can turn into a three-year integration project. Transactions happen in the streaming stack, with processing measured in milliseconds. But to know if there is fraud, the bank needs historical patterns from its analytics stack, batch-processed nightly. The ML model is trained in the ML stack, requiring data in specific formats, which neither streaming nor analytics can provide. Business rules—spending limits, merchant categories, customer segments live in the engineering stack, accessible only through ETL jobs. Making a single fraud decision requires orchestrating across all four stacks in milliseconds. Miss one connection and either fraud slips through, or legitimate customers get blocked at checkout. This is how four-stack complexity turns a simple use case into months of integration work and millions in preventable losses.

In such cases, attempts at integration create their own problems. Each "solution" adds layers of complexity that compound the original problem. Teams build point-to-point connections: streaming to ML, warehouse to ML, warehouse to streaming. Each connection is brittle, breaking whenever a system changes. The result is architectural spaghetti; changing anything risks breaking everything.

Worse still, each stack has its own version of truth. The streaming stack shows 50,000 active customers (real-time sessions), the warehouse shows 45,000 (last night's batch), and the ML stack shows 52,000 (including predicted churn-backs). Which should be used for AI training? Teams cloud spend months reconciling these differences instead of building new capabilities.

From Data Swamps to Lakehouse

The four-stack problem is only half the story. The storage layer evolved in its own fragmented way, creating a second source of complexity. In the 1990s, we built data warehouses—structured, organized, efficient. They were like perfectly planned cities with designated zones for everything. But they couldn't handle anything unexpected. Try to store customer sentiment or social media data in a warehouse designed for transaction records, and things go sideways.

By the early 2010s, the pendulum had swung to data lakes. The promise was seductive. "Store everything! Raw data, structured data, unstructured data. Dump it all in one place and sort it out later." The freedom was intoxicating. But data lakes soon became data swamps: unorganized repositories where no one could find what they needed.

In 2020, the lakehouse emerged—an architecture that elegantly balanced structure with flexibility. Think of a lakehouse as combining the best of planned infrastructure with organic growth, like structured zones for critical operations, flexible spaces for innovation, and clear paths connecting everything.

A lakehouse provides the performance and reliability of warehouses with the flexibility and scale of lakes. Instead of choosing between structure and flexibility, both are supplied. Instead of moving data between systems, process it where it lives. Instead of maintaining separate copies for different uses, only one source of truth—accessible in multiple ways—is maintained.

Lakehouse architecture is protection from paying twice, once to store data in a warehouse for reports, then again to store it in a lake for AI. Teams don't waste months moving data between systems. When the CMO wants to predict customer behavior, they can access transaction data, social media sentiment, and customer service logs from one place. Speed to insight improves dramatically.

Uber's lakehouse approach cut data latency from a full day to under an hour.[10] Netflix reported 60 percent query performance improvements and 70 percent reduction in compute instances through their lakehouse data optimization.[11]

A lakehouse uses open formats like Delta Lake, Apache Iceberg, or Apache Hudi. These provide ACID transactions, schema enforcement, and time-travel capabilities while supporting structured and unstructured data. Major platforms like Databricks, Snowflake, and various cloud providers offer production-ready implementations. Platforms that support open formats provide a migration path. While moving data between vendors may still be costly (e.g. data egress fees), having that option is far better than being a vendor locked in with no escape route.

The Modern Data Foundation

Beyond choosing the right platform, successful lakehouse implementations follow proven patterns.

The Medallion Architecture

A way of organizing data quality in three progressive layers:

- **Bronze layer**: Takes raw data exactly as it arrives, messy, duplicated, but complete. This is the safety net when someone asks, "What did the source system actually send?"

- **Silver layer**: Cleaned, deduplicated, and conformed data, where mismatches and standardized customer IDs are fixed.

- **Gold layer**: Business-ready datasets optimized for specific use cases. The "customer 360" lives here (combining purchase history, support tickets, and preferences into a single profile).

A retailer's journey through layers included bronze showing 5 million customer records, silver revealing 3.8 million actual customers after deduplication, and gold producing 2.1 million active customers with complete profiles ready for AI. Each layer has value: bronze for audit trails, silver for exploration, gold for production AI.

Beyond this layering architecture, these practices will maximize efficiency:

Domain Organization

This prevents chaos at scale. Instead of dumping everything into one bucket, organize using business domains that teams understand, like customer data, product data, financial data. Organize a lakehouse the way the business thinks, not the way IT does.

Storage Tiers

These directly impact costs.

- **Hot storage (most expensive)**: Keeps frequently accessed data instantly available, like the last 90 days of transactions.

- **Warm storage (moderate cost)**: Balances price and access for occasional queries, like the last two years of history.

- **Cold storage (cheapest)**: An archive of compliance data that hopefully will never be touched (e.g. anything beyond two years).

The key is to automate these transitions based on access patterns so that data ages gracefully without manual intervention.

Built-in Security

Critical to lakehouse success is built-in security. Here's how the four principles I covered early in this chapter translate into lakehouse-specific practices. Your security team will implement additional layers (e.g., key management, privileged access controls, egress filtering), but these nine practices are the non-negotiable foundation. Verify they're in place.

Encryption

- **At rest**: AES-256 or similar standards for all stored data.

- **In transit**: TLS 1.2 or higher for all data movement.

Access Control

- **Identity**: Integrate with a central identity provider; multi-factor authentication is mandatory, not optional.

- **Role-based access control (RBAC)**: Map users to functional roles, not individual permissions.

- **Fine-grained permissions**: Column-level and row-level security let analysts query purchase patterns without ever seeing underlying payment data.

Containment

- **Network segmentation**: Isolate data domains so a breach in one area cannot spread laterally. Marketing analytics should have no path to payment processing.

- **Environment separation**: Maintain strict boundaries between development, staging, and production, credentials that work in dev should never work in production.

Audit

- **Immutable logs**: Track every query, access, and modification, stored in a separate, locked-down environment where even data team members cannot delete records.

- **Anomaly detection**: Passive logging isn't enough. Automated alerts should flag bulk data exports, logins from unexpected locations, and unusual query patterns.

Structure

Modern lakehouse architectures also solve the unstructured data challenge. Remember, 90 percent of enterprise data is unstructured. Document stores handle text-based content, while vector databases store embeddings (numerical representations of text or images created by AI models), which enable similarity searches based on meaning rather than keywords.

Unified query engines are a way to analyze everything together. A manufacturer can now combine sensor data from time series, maintenance photos as images, technician notes as unstructured text, and equipment specifications as structured data. AI can predict equipment failures days in advance by finding patterns across all data types, something that was very difficult and costly to achieve with traditional architectures.

How Modern Architecture Solves the Four-Stack Complexity

Lakehouse architecture elegantly collapses the four-stack complexity into one unified platform. Here's how each problematic stack gets absorbed and improved:

- **Unified analytics and streaming**: Instead of separate systems for "what happened" and "what's happening," Lakehouse handles both. Fraud detection model trains on five years of historical transactions *and* scores real-time payments from the same data platform. No more shuttling data between incompatible systems. Delta Lake, Apache Iceberg, and similar technologies provide ACID transactions (guaranteed data reliability and consistency) on streaming data, which traditional architectures cannot deliver.

- **ML-native processing**: Rather than requiring separate ML platforms that need special data formats, lakehouse platforms incorporate ML capabilities natively. Data scientists work directly with production data. No more training on stale exports that don't match reality. Feature engineering happens once and serves both training and inference. The same SQL query that generates a report can feed an ML model. Databricks, Snowflake, and cloud platforms now include built-in ML runtimes, eliminating the translation layer between analytics and AI.

- **Business logic as code**: The 10,000 SQL scripts scattered across an engineering stack become version-controlled transformation pipelines in Lakehouse. Using tools like dbt (data build tool) or native transformation capabilities, business rules become testable, documented, reusable assets. When marketing changes the definition of "high-value customer," it propagates automatically to every downstream use case.

However, architecture alone isn't enough. Orchestration, the intelligent coordination that makes data flow seamlessly, is needed. Platforms like Apache Airflow, Dagster, or Prefect act as the conductor of the data orchestra. They ensure:

- Data arrives from source systems on schedule (or triggers on events).

- Transformations execute in the correct sequence.

- Quality checks run before data moves downstream.

- Failed jobs retry intelligently without corrupting data.

- Dependencies are respected, with customer profiles updated before personalization models are trained.

Think of orchestration as a data supply chain manager. When an order is entered through the streaming system, orchestration ensures it flows through fraud detection (real-time), updates inventory (near real-time), appears in analytics (within minutes), and triggers replenishment if needed (event-driven). All automatically, monitored, and with clear error handling.

A complete solution stack looks like this:

- **Data storage and processing**: Lakehouse platforms provide unified storage, SQL analytics, streaming processing, and ML capabilities in one platform. Available solutions include Databricks, Microsoft Fabric, Google BigQuery with BigLake, Dremio, Starburst, and other cloud-native platforms.

- **Orchestration layer**: Modern workflow orchestration coordinating data movement, enforcing dependencies, handling errors, and ensuring data freshness. Tools in this category include Apache Airflow, Dagster, Prefect, Astronomer, Azure Data Factory, AWS Step Functions, Google Cloud Composer, Luigi, and Temporal.

- **Transformation layer**: Code-based transformation tools turning business logic into maintainable, version-controlled assets. Options include dbt, Apache Spark, Apache Flink, Apache Beam, Matillion, native platform capabilities, and SQL-based transformations.

- **Governance layer**: Data governance platforms providing unified security, lineage, and quality management across all data and use cases. Solutions include Unity Catalog, Microsoft Purview, AWS DataZone, Google Dataplex, Collibra, Alation, Atlan, and similar platforms.

The data transformation extends beyond technology. When four stacks are collapsed into one, teams become more unified. Data engineers, analysts, and scientists work on the same platform using familiar tools. The warehouse team's SQL expertise applies to streaming data; the streaming team's real-time skills enhance analytics; the ML team's models deploy without translation layers. Knowledge compounds instead of fragmenting.

From Architecture to Organization

For large, complex organizations, building a well-architected lakehouse is challenging. Not technically, but organizationally. When a central team owns all the data of a 50,000-person company, it becomes the bottleneck. For example, understanding what "active customer" means to marketing versus finance versus operations is a huge lift, as marketing might define it as "opened an email in the last 30 days," while finance might define it as "made a purchase in the last 90 days". It's hard to move fast enough to serve everyone.

Data mesh offers a solution, transforming how organizations think about data ownership while maintaining the shared lakehouse infrastructure, which will provide a unified technical foundation. Data mesh adds the organizational layer that determines who owns, maintains, and publishes data products within the shared lakehouse.

Domain teams own their data end to end, marketing customer data, finance revenue data, and operations supply chain data. Each domain treats its data as a product, with clear owners, quality guarantees, and published interfaces for others to consume. This transforms data from a centralized bottleneck into distributed assets. Think of it like a shopping mall: shared infrastructure like parking, security, and HVAC, but each store manages its own inventory and customer experience.

Making Data Products Easy to Consume with Data Contracts

With data mesh, each department treats its data as a product published for others to consume, covered by data contracts. Think of it like API contracts, but for data.

Just like APIs define how applications talk to each other (what endpoints exist, what parameters they accept, what data they return), data contracts define how teams share data products. While traditional APIs serve real-time requests, data contracts also cover batch data, streaming feeds, and analytical datasets. They specify what's available, computational requirements, included fields, and access restrictions. Data producers can't change their data products without versioning and migration paths. Breaking changes require notice and migration support.

When a department needs to change something, they version it, maintain both versions for the notice period, and help consumers migrate. This requires formal data contracts with clear specifications and change notice requirements. The best data contract is one that's simple enough to understand and formal enough to rely on.

Data Mesh Pitfalls: When Domains Collide

When domains don't coordinate, conflicting definitions return. Critical metrics like "delivery time" or "incident severity" mean different things across teams. AI learns all versions as valid, making inconsistent decisions about the same operational scenarios. Without clear governance, ownership disputes can paralyze progress.

The solution is to establish clear ownership and accountability from day one. For any metric that drives decisions, one domain owns the golden source, while others can enhance but not contradict. When data serves multiple departments, those departments share accountability for data quality. Infrastructure costs need to be linked to value creation, with departments benefiting from data products contributing to their cost. Such a governance model transforms organizational velocity.

Instead of filing access requests, users get role-based permissions, and rather than central control, domains have federated responsibility. The result is business teams that can make data-driven decisions without IT bottlenecks. Analysts can spend time on analysis rather than access requests, and departments stop building their own shadow databases (fewer Excel workarounds).

Making Data Findable

Searching for data is harder than recreating it. Teams can't use data they can't find, so they build shadow copies. Every shadow copy drifts from the original, and now there are multiple versions of truth despite the architecture being designed to prevent exactly that.

A data catalog is the fix. It indexes every dataset across the organization, documents what each contains, who owns it, how fresh it is, and how it connects to other datasets. That last part, lineage, matters specifically for AI. When a model makes a decision that a regulator questions, the answer requires tracing from the model's output back through every transformation to the original source data. Without lineage, that question takes weeks. With it, it takes minutes.

The catalog also closes the loop on data mesh. Domain teams publish data products, but consumers need a way to browse what's available, understand what it means, and assess whether it's fit for their purpose, all without filing a ticket or scheduling a meeting with the owning team.

The Investment Reality

Enterprise-wide data transformations require significant investment. Leading companies across industries are making this clear. JPMorgan Chase, for example, highlights in its 2023 Annual Report a $2 billion build-out of new data centers, alongside long-term commitments to modern data platforms and AI-ready infrastructure.[12] Hidden costs that catch executives off guard need to be budgeted for. For example, cloud egress fees for moving data out of a cloud provider can add thousands monthly. Private connectivity on public clouds adds extra fees that compound quickly across multiple services. Parallel system costs during transition can double infrastructure spend. To uncover all hidden costs, run realistic pilots (I'll cover that in Strategy 7). The point is that building a modern data foundation is expensive. Most end up underfunded because they are treated as an IT line item. Treat yours as strategic infrastructure, like building a factory or acquiring a competitor. The investment required, even with hidden costs, is a fraction of what you'll pay for AI built on bad data.

Case Study: Uber's 1.5 Exabytes

Imagine asking 10,000 people to share the same refrigerators and every time they want something, they have to wait 24 hours for it to defrost. That's what Uber's data infrastructure felt like before its transformation.

Today, Uber's workforce includes more than 10,000 data professionals—from scientists analyzing patterns to operations teams managing cities to analysts forecasting business trends to engineers building systems—all accessing the same pool of 1.5 exabytes stored across two facilities.[13] Every single day, this generates more than 500,000 database queries and 370,000 analytical applications.[13]

The scale created problems that more hardware alone couldn't solve. As data volume grew and users created their own analysis jobs, the system accumulated countless small files that strained the infrastructure.[10] The NameNodes—the directory systems managing file locations—faced mounting load under these conditions.[10] The system needed a full 24-hour cycle before newly generated data became available to anyone.[10]

For a business operating in real-time, yesterday's numbers are worthless.

Most organizations facing this problem would throw money at it. Migrate to the cloud. Scale up storage. Add computing power. Uber did migrate to the cloud, but they understood that cloud scale without architectural redesign can make the problem worse, not better. The bottleneck wasn't technical capacity. It was organizational structure.

Uber created a service called DataMesh that automated something most companies do manually: figuring out who owns what data.[13] The company already tracked which teams were responsible for which parts of the business in an internal system called uOwn.[13] DataMesh used that information to assign each dataset to the appropriate team and structure the cloud resources to match.[13] No committees. No approval processes. Just automated mapping based on how the business actually operated.

What changed wasn't just where the data lived. It changed who owned it.

Instead of 10,000 people depending on a central team to understand every use case and prioritize every request, domain teams became responsible for their own data. The infrastructure still provided shared services—security, governance, access controls—but ownership was distributed to the people who actually understood what the data meant and how it should be used.

The performance numbers validated the approach. What previously took an entire day to surface now arrived in thirty minutes for raw data.[10]

Most companies eventually hit the same wall: centralized data management can't keep pace with decentralized data needs. The instinct is to add more capacity. The solution is to redesign the organization. The question isn't just "how much data do we have." It's "can people actually find, understand, and use it."

STRATEGY 3 ACTION PLAN

Days 1-30: This week, schedule a meeting with the head of IT and the highest-revenue business unit leader. Ask one question. "Who decides what customer data means and who can use it?" If IT answers "we do" while the business leader looks confused, the main problem has been identified. Data decisions aren't IT's to make; business leaders must define what the data means, who uses it, and what quality is acceptable. IT implements these business decisions, ensuring security and governance to maintain compliance. I.e. business owns the "what" and "why" of data. IT owns the "how."

Pick the most critical data domain, likely customer, product, or financial data. Assign a business executive as the owner. This person now owns data quality, access, and definitions for their domain.

Map current technology stacks. Identify integration gaps between analytics, streaming, engineering, and ML systems.

Run a quality baseline on the pilot domain. How many duplicate customer records? How many missing fields? What percentage of data is actually usable? Set a baseline on which improvement can be measured.

Days 31-60: Keep legacy systems running and build a modern data platform alongside, extracting data through ETL pipelines into the lakehouse bronze layer. Migration can wait, integration can't.

Default to managed data platforms, they handle complexity that's easy to underestimate. Open-source offers more control, but only delivers value if your team has genuine technical depth, not just confidence. Go open source only if you have a very specific need not met by managed services. Don't chase perfection; speed matters.

Start with one domain and one use case. Implement medallion architecture—bronze, silver, gold—for just this domain. Build security from day one with encryption, access controls, security boundaries between systems, and audit logs. Add domains only after the first proves value.

Configure storage tiers: hot for frequently accessed data, warm for occasional queries, cold for compliance archives.

Create the first data contract between two business domains specifying what's shared, when it updates, who owns it. One domain owns the authoritative version. No committees, one owner. Configure a data catalog to index all datasets in the pilot domain.

Budget follows ownership. Does marketing own customer data? Then marketing pays for it.

Days 61-90: Include data ownership in performance reviews. Measure progress against improvement in data quality from baseline. Transformation sticks when quality becomes a business KPI, not an IT metric.

By day 90, you should see business leaders negotiating data contracts directly with each other, at least 50 percent reduction in the time needed to get data for the pilot domain, and at least one AI initiative previously stalled on data now moving forward.

1 Sean Steinberg, Adam Stepan, and Kyle Neary. "Target Cyber Attack: A Columbia University Case Study." Columbia University School of International and Public Affairs. 2021. https://www.sipa.columbia.edu/sites/default/files/2022-11/Target%20Final.pdf

2 "What We Learned from Target's Data Breach 2013." CardConnect. March 19, 2023. https://cardconnect.com/launchpointe/payment-trends/target-data-breach/

3 Andrew McAfee and Erik Brynjolfsson. "Big Data: The Management Revolution." *Harvard Business Review*. October 2012. https://hbr.org/2012/10/big-data-the-management-revolution

4 Randy Bean and Thomas H. Davenport. "2024 Data and AI Leadership Executive Survey." Wavestone. January 2024. https://www.wavestone.com/wp-content/uploads/2024/11/data-ai-executive-leadership-survey-2024.pdf

5 Nicolaus Henke, Jacques Bughin, Michael Chui, James Manyika, Tamim Saleh, Bill Wiseman, and Guru Sethupathy. "The Age of Analytics: Competing in a Data-Driven World." McKinsey Global Institute. December 2016. https://www.mckinsey.com/capabilities/quantumblack/our-insights/the-age-of-analytics-competing-in-a-data-driven-world

6 "How Much Does It Cost to Develop an AI Solution? Pricing and ROI Explained." Coherent Solutions. Last updated November 24, 2025. https://www.coherentsolutions.com/insights/ai-development-cost-estimation-pricing-structure-roi

7 Thomas C. Redman. "Bad Data Costs the U.S. $3 Trillion Per Year." *Harvard Business Review*. September 22, 2016. https://hbr.org/2016/09/bad-data-costs-the-u-s-3-trillion-per-year

8 "Data Quality: Why It Matters and How to Achieve It." Gartner. https://www.gartner.com/en/data-analytics/topics/data-quality

9 Thomas C. Redman. "Data's Credibility Problem." *Harvard Business Review*. December 2013. https://hbr.org/2013/12/datas-credibility-problem

10 "Uber's Big Data Platform: 100+ Petabytes with Minute Latency." Uber Blog. October 17, 2018. https://www.uber.com/blog/uber-big-data-platform/

11 Netflix Technology Blog. "Optimizing Data Warehouse Storage." Medium. December 21, 2020. https://netflixtechblog.com/optimizing-data-warehouse-storage-7b94a48fdcbe

12 "Annual Report 2023." JPMorganChase. https://www.jpmorganchase.com/ir/annual-report

13 Uber Engineering. "DataMesh: How Uber Laid the Foundations for the Data Lake Cloud Migration." Uber Blog. September 10, 2024. https://www.uber.com/blog/datamesh/

Build from Within, Hire to Multiply

| Your Best AI Talent Already Works for You

Organizations spend millions annually on AI talent acquisition. Six months later, these brilliant minds are doing work junior analysts could handle, or they've already left for companies where they can do "real AI work."

Meanwhile, internal champions who've been delivering real value watch external hires get resources they're denied. The truth is that the best AI talent already works for you. They just need to be integrated, equipped, and trained, not replaced.

These internal hidden champions possess irreplaceable assets, including a deep understanding of business context, established relationships that accelerate adoption, knowledge of what actually matters to customers, and long-term commitment to success. A McKinsey survey of executives found that only 57 percent of organizations plan to build their GenAI capabilities predominantly through upskilling and redeploying existing talent rather than external hiring.[1]

Organizations fall into the credential trap, hiring algorithm-first thinkers instead of outcome-focused problem solvers. The problem is that people who speak Python don't necessarily speak business, and they are working alongside people who understand business but can't leverage AI.

Current market data shows ML engineers at tech companies command packages exceeding $300,000.[2] But matching compensation isn't enough. Tech companies don't just pay more; they offer meaningful problems, cutting-edge tools, and visible impact. When data scientists are stuck cleaning data and building PowerPoints, no salary premium will retain them. When you do identify skill gaps that require external hires, create the environment—meaningful problems, real empowerment—that makes them stay.

The Human Barriers

Getting the right people in place is half the challenge. As important is human dynamics that can slow momentum. Three patterns surface repeatedly, not necessary from bad intentions, but because people instinctively protect what they've built:

1. AI teams operate in isolation. IT sees them as rogue developers bypassing security protocols. Business units see them as a corporate tax eating into their budgets. Finance can't measure its ROI with traditional metrics. With no natural allies and everyone protecting their territory, isolated AI teams fight battles on all fronts instead of building solutions. Don't build isolated AI teams.

2. Managers see AI teams as a threat to their expertise. They control resource allocation and adoption within their teams. When they resist with slow approvals, endless requirements, or passive aggression, AI initiatives can die by a thousand cuts. Make managers sponsors of AI initiatives. Turn them into catalysts, not obstacles to overcome.

3. AI teams become dumping grounds for every technology idea. "Since you're doing AI, can you look at blockchain?" and "The board mentioned quantum computing." Instead of creating value, they're stuck chasing every shiny new technology. Define roles clearly. AI teams work on specific measurable initiatives, not emerging tech exploration.

The 60–30–10 Formula

Building an effective AI team isn't about hiring the smartest people; it's about combining the right capabilities. The optimal mix tends to be 60-30-10. Sixty percent internal talent, 30 percent specialized experts who can multiply capabilities, and 10 percent strategic partners for time-bound knowledge transfer. This solves the isolation problem discussed earlier while still maintaining a unified platform. The central AI team (the 30 percent) handles standards and infrastructure, while the 60 percent splits into two groups: translators who align AI with strategy across departments, and domain experts who apply AI within their specific functions.

Domain experts make up the bulk of that 60 percent. You only need one or two translators per project and they can serve multiple projects simultaneously because they work horizontally across the organization. The success of your AI projects depends on the translators more than any other group.

The 60-30-10 formula establishes AI capability across the organization. Once that foundation is in place, you can adjust the ratios as needed for specific projects.

The 60 Percent: An Internal Transformation Force

Sixty percent of the team should come from existing employees who deeply understand the business. Smart organizations prioritize internal talent development because no one understands how to execute and sustain change better than your own people.

Business Translators: The true MVPs

The first part of the 60 percent are business translators who can determine whether AI investments generate returns or are just nice models. McKinsey's research identifies them as playing "a critical role in bridging the technical expertise of data engineers and data scientists with the operational expertise" of business functions.[3]

The most successful translators share three traits that can't be taught in a classroom.

- **Influence without authority**: Both engineers and executives listen to them because of their track record, not their title.

- **Systems thinking**: They see connections others miss, linking data from one function to solve problems in another.

- **Productive impatience**: They're frustrated by inefficiency but channel it into prototyping solutions, not just complaining.

Business translators know which edge cases could destroy value. A data scientist builds perfect customer segmentation, but a translator knows that treating high-value customers based on algorithmic clustering alone could be catastrophic. When translators say 'this will help you hit targets,' their sponsors believe them, because they've demonstrated they understand what those targets mean.

Translators' main role is to turn technical possibilities ('predict churn with 85 percent accuracy') into strategic opportunities ('save $10 million annually retaining high-value customers'). They facilitate sessions between data scientists and domain experts, create business-friendly visualizations of model outputs, design feedback loops based on business outcomes (not just technical metrics), and calculate ROI in CFO-friendly terms.

To find them, look for employees with learning agility who've successfully adapted to new tools before. Those who influence without authority. People that others naturally consult for advice. People who have led successful process improvements, who constantly ask 'why' and 'what if' questions, and who have volunteered for stretch assignments.

Domain Experts with AI Skills

The second part of the 60 percent are domain experts who became AI practitioners in their specific areas, such as supply chain analysts and marketing managers. These are the people who know exactly where value hides and risks lurk.

Transform them through targeted enablement. Six-week intensive training in practical AI tools, not neural network theory, but "How to use AutoML for fraud detection." Provide them access to platforms with guardrails that prevent common errors, and mentorship from architects and MLOps engineers who act as coaches, not doers. Most critically, give them protected time to work, at least 50 percent on AI projects without abandoning regular duties.

These domain experts deliver value that external hires never could. They understand every edge case that could cost millions if handled incorrectly. They know which regulations matter and which customer quirks determine success. They have relationships that accelerate adoption.

The 30 Percent: Strategic External Catalysts

Thirty percent should be strategic external hires. If they've already been brought in, verify they're positioned correctly. These aren't individual contributors; they're force multipliers who make everyone else more effective. Focus on two critical roles:

- **AI solution architects**: They design scalable, secure infrastructure that others build upon. They don't create every model; they create platforms enabling the 60 percent to build solutions safely and efficiently. Their role is building the right abstractions, creating reusable components, establishing security guardrails that prevent both errors and breaches, embedding governance controls from day one, and designing systems that scale from prototype to production without compromising compliance. They ensure every model built on the platform meets security standards and regulatory requirements by default, not as an afterthought. They're expensive but worth it because they multiply everyone's effectiveness while protecting from catastrophic failures.

- **MLOps engineers**: They establish the infrastructure and processes that keep AI systems reliable and compliant at scale, such as pipelines that automate model training, testing, and deployment with validation checks at every stage; monitoring systems that detect model drift, performance degradation, and anomalous behavior; audit trails that satisfy regulators and governance frameworks that ensure reproducibility; and access controls determining who can deploy models and implement circuit breakers for when models misbehave. Without MLOps engineers, you're not just flying blind but also without a black box recorder to hand over to regulators who come asking why the model made a particular decision.

The 10 Percent: Knowledge Partnerships

Specialized partners providing expertise for time-bound challenges are the last part of the team. This isn't staff augmentation by which generic resources are rented; it's surgical expertise for very specific problems where building permanent capability doesn't make sense.

The best partners share three characteristics. They transfer knowledge while solving problems by pairing with your team, gradually transferring ownership, and explicitly planning their exit. They bring specialized expertise that is occasionally needed, like computer vision for a manufacturing quality project or NLP for customer service automation. Finally, they measure success by how quickly they make themselves unnecessary.

For example, a computer vision expert is engaged for three months to build quality inspection models for manufacturing. But they don't just deliver models. They train your domain experts to maintain and improve them, documenting the approach so it can be replicated, and build templates that can be adapted for other use cases. The expertise is rented, but the knowledge is purchased.

Avoid partners who create dependencies, like those who offer black box solutions you can't understand or modify, or proprietary contracts without knowledge transfer. The goal is building in-house capability, not permanent reliance on external expertise.

Executive Sponsorship: Beyond Budget Approval

Buy-in (leadership approval and budget) isn't enough. Sponsorship (active engagement throughout) is what drives AI success. We touched on that already in Strategy 2. In McKinsey's study of AI implementation leaders, "77 percent of ML implementation leaders had C-level leadership driving their projects. For 44 percent of leaders, digital projects were sponsored by the CEO or board of directors—more than double the rate of companies in the bottom 50 percent of the sample."[4]

Effective executive sponsors play three distinct roles. As vision setters, they connect AI to strategy, answering the "why" that motivates teams and the "so what" that engages stakeholders. As obstacle removers, they cut through red tape that teams can't overcome alone, like data silos protected by politics, risk-averse legal teams blocking experimentation, or IT policies preventing AI adoption. As success amplifiers, they shape organizational perception through what they celebrate and how they respond to failures.

You're ready for a CAIO when there are multiple AI initiatives across business units that require coordination, big AI spending that requires strategic allocation, and regulatory exposure demanding enterprise-wide governance. The successful CAIO connects AI to business strategy, builds capability across the organization, and manages governance. They're an orchestrator, not a super data scientist.

If the above criteria aren't met, don't force it. Establish an AI Council composed of your business translators, each owning AI initiatives within their domain. They coordinate strategy and share learnings, but each maintains single-threaded accountability for their own initiatives. This builds capability while identifying who could become a future CAIO, a role that often emerges from business leadership, not technical roles.

Empowering Your Team with AI

Give teams AI tools designed for their actual day-to-day work, while embedding AI fluency into daily operations. This isn't about making everyone an AI expert; it's about changing how they get work done.

Research on human-AI collaboration reveals that the greatest performance improvements come not from replacing humans with AI, but from augmenting human capabilities with AI tools.[5] GitHub's research shows developers using AI assistance complete their tasks 55 percent faster,[6] with reduced cognitive load on routine tasks and more time working on architecture than syntax. The goal is 10x productivity improvement, not 10 percent efficiency gain. That is to say, don't measure AI productivity by the time an employee saves (e.g., saved 30 minutes a day because they no longer write every email from scratch) but rather by what that employee is now doing with those 30 minutes they saved.

Tools alone don't create transformation. People need permission and time to experiment with AI tools without fear of looking incompetent. Start with familiar applications like AI meeting summaries, automated report generation, or intelligent document search. Gradually introduce more sophisticated capabilities as teams build confidence. Most importantly, embed AI into existing workflows rather than creating separate "AI projects." The transformation happens when AI becomes how people work, not something extra.

Case Study: JPMorgan Chase

JPMorgan Chase faced the same choice every large enterprise confronts: compete with Google and Microsoft for AI talent, or find another way.

The bank chose to bet on the 300,000 employees who already understood banking, risk, compliance, and customers.[7]

That bet paid off. By 2023, JPMorgan had over 300 AI use cases running in production across risk management, fraud detection, marketing, and customer experience.[8] These were actual production systems, not pilots. The bank achieved this not by building an isolated AI lab, but by embedding AI capability across the organization.

They understood they didn't need to turn everyone into a data scientist. The goal was clear: AI had to change how people worked. JPMorgan launched "AI Made Easy," a training program designed for employees with no technical background. Tens of thousands completed it.[9] The bank also made AI training mandatory for new hires, with prompt engineering now part of onboarding.[10] These are domain experts gaining AI skills in their specific areas.

JPMorgan's internal AI platform, LLM Suite, now reaches nearly a quarter-million employees.[9] As Derek Waldron, JPMorgan's chief analytics officer, described, "Lawyers use it to scan, read, compare, and generate contracts; credit professionals use it to read terms, compare covenants, and extract information; sales professionals and frontline bankers use it to distill information and prepare for meetings."[99] Each group applies AI to problems they understand deeply.

One early example demonstrated the power of domain expertise combined with AI. The bank's COIN system automated review of commercial loan agreements, work that previously consumed 360,000 hours of lawyer and loan officer time annually.[11] The people who understood those contracts best were now freed to focus on judgment calls, not pattern recognition.

JPMorgan was deliberate about external talent. In 2018, the bank brought in Manuela Veloso from Carnegie Mellon's Machine Learning Department to lead AI research.[12] The partnership with Carnegie Mellon created an AI Maker Space where researchers work on synthetic financial datasets.[13] These external investments act as force multipliers, transferring knowledge that thousands of internal practitioners can learn from.

The bank's $18 billion annual technology budget gets the headlines.[9] But the real transformation isn't about spending, it's about where capability lives. As Waldron emphasizes, JPMorgan's progress came from treating AI as a cultural transformation, not a tooling exercise. Leaders had to rethink how work gets done, build new skills across the organization, and enable cross-functional teams to apply AI in their day-to-day decisions.[9]

JPMorgan didn't chase unicorns. It transformed the experts it already had.

STRATEGY 4 ACTION PLAN

Days 1-30: This week, identify potential domain experts by looking for Excel power users pushing spreadsheets to their limits, process innovators who constantly improve workflows, and subject matter experts who others consult for domain-specific questions. Survey for employees who've completed online AI courses on their own time; they're showing initiative.

Assess executive engagement levels. Who would make an effective AI sponsor? Who has the capacity and influence to drive initiatives forward?

Create a list of 10–20 internal domain expert candidates and three to five potential executive sponsors.

Present the AI opportunity to executives framed around their strategic priorities. Get specific sponsorship commitments, like eight hours weekly, direct obstacle removal, and public success amplification.

Select two to three potential business translators. Test their ability to bridge technical and business worlds through practical exercises. Can they explain a technical concept to sales and a business need to IT? Can they translate a technical capability into business impact?

Define the team following the 60–30–10 formula. Identify which internal talent to develop, which external roles are critically needed, and partnerships that make sense. For the partnership 10 percent, identify time-bound expertise needs where building permanent capability doesn't make sense.

Deploy AI tools that match the team's day-to-day work. Start simple—meeting summaries, email drafting, research assistance— and build from there. Give people permission to experiment. Measure success by what they accomplish with the time saved, not the time itself.

Days 31-60: Create a small AI Center of Excellence that reports to the CIO. This is where the AI architect(s) and MLOps engineer(s) sit permanently. Their job isn't to build every solution; it's to establish standards, governance, and reusable infrastructure that others build upon.

Deploy this technical core to the first business problem. Pull in three or four domain experts from the affected department who join for this specific mission while keeping their day jobs. The business translator leads the team. The domain experts rotate based on the problem being solved; the technical core remains constant and provides continuity across projects.

This hub-and-spoke model gives you centralized governance without creating an isolated AI team. Technical expertise lives in one place for standards and oversight; execution happens embedded in the business.

Connect the mission directly to strategic priorities with measurable success metrics. Watch for middle managers who see AI as a threat to their expertise. Address resistance early.

Days 61-90: Based on early success, identify the next wave of internal domain experts. Deploy the AI Center of Excellence to a second business problem in a different area; the technical core now supports two projects simultaneously while new domain experts rotate in. Your translator can support both projects, working across them to maintain strategic alignment. Have the first team's domain experts mentor the new ones. Target two to three pilots in production by day 90. Measure success by whether business units actively request AI help. Let demand drive growth of the AI Center of Excellence.

---

[1] Asin Tavakoli, Henning Soller, and Suman Thareja. "Delivering the Right Data Talent for Your Data Transformation." McKinsey & Company. July 30, 2025. https://mckinsey.com/capabilities/mckinsey-digital/our-insights/tech-forward/delivering-the-right-data-talent-for-your-data-transformation

[2] Sims, Andy. "How Much Do Machine Learning Engineers Make? (2025 Guide for HR & Compensation Teams)." SalaryCube. https://www.salarycube.com/compensation/average-machine-learning-engineer-salary/

[3] Nicolaus Henke, Jordan Levine, and Paul McInerney. "Analytics Translator: The New Must-Have Role." McKinsey & Company. February 1, 2018. https://mckinsey.com/capabilities/quantumblack/our-insights/analytics-translator

[4] Arnav Dey, Bruce Lawler, Delphine Zurkiya, Vijay D'Silva, and Vivek Arora (with Kyle Danner). "Bold Accelerators: How Operations Leaders Are Pulling Ahead Using AI." McKinsey & Company. August 19, 2025. https://mckinsey.com/capabilities/operations/our-insights/mind-the-gap-how-operations-leaders-are-pulling-ahead-using-ai

[5] James Wilson and Paul R. Daugherty. "Collaborative Intelligence: Humans and AI Are Joining Forces." *Harvard Business Review*. July-August 2018. https://hbr.org/2018/07/collaborative-intelligence-humans-and-ai-are-joining-forces

[6] Eirini Kalliamvakou. "Research: Quantifying GitHub Copilot's Impact on Developer Productivity and Happiness." GitHub Blog. September 7, 2022. https://github.blog/2022-09-07-research-quantifying-github-copilots-impact-on-developer-productivity-and-happiness/

[7] "JPMorgan Chase Rolls Out Firmwide AI Training for 300,000 Employees." IndexBox. October 31, 2025. https://www.indexbox.io/blog/jpmorgan-chase-rolls-out-firmwide-ai-training-for-300000-employees/

[8] Ross Henry Law. "J.P. Morgan Reveals It Has 300 AI Use Cases in Production." FStech. April 12, 2023. https://www.fstech.co.uk/fst/JP_Morgan_reveals_it_has_300_AI_use_cases_in_production.php

[9] "JPMorgan Chase's Derek Waldron on Building an AI-First Bank Culture." McKinsey & Company. October 29, 2025. https://www.mckinsey.com/industries/financial-services/our-insights/jpmorgan-chases-derek-waldron-on-building-an-ai-first-bank-culture

[10] "JPMorgan: All New Employees Will Receive AI Training." PYMNTS. May 20, 2024. https://www.pymnts.com/news/banking/2024/jpmorgan-all-new-employees-will-receive-ai-training/

[11] Debra Cassens Weiss. "JPMorgan Chase Uses Tech to Save 360,000 Hours of Annual Work by Lawyers and Loan Officers." ABA Journal. March 2, 2017. https://www.abajournal.com/news/article/jpmorgan_chase_uses_tech_to_save_360000_hours_of_annual_work_by_lawyers_and

[12] Nathan DiCamillo. "Carnegie Mellon Professor to Lead JPMorgan Chase AI Research." American Banker. May 9, 2018. https://www.americanbanker.com/news/carnegie-mellon-professor-to-lead-jpmorgan-chase-ai-research

[13] Aaron Aupperlee. "CMU Invites Students to Explore Artificial Intelligence With Opening of JPMorgan Chase & Co. AI Maker Space." Carnegie Mellon University. November 10, 2021. https://www.cmu.edu/news/stories/archives/2021/november/jp-morgan-ai-maker-space.html

From Fifty Vendors to the Five That Matter

| If It Can't Comply, It Doesn't Earn Its Place

Picture a Fortune 500 firm, flush with AI ambition and a healthy budget. It has assembled what is believed to be a best-of-breed AI stack, utilizing ten vendors. Each promised to be "the solution."

It already had an AI platform and tools that worked fine. But every vendor pitch made the existing infrastructure sound obsolete, so it bought new "everything."

Fast forward three years. The CTO is staring at a monster of their own making, a Frankenstein ecosystem featuring tools that speak different languages, data lives in silos, and the integration team has tripled in size just to keep the whole thing from falling apart. Their "state-of-the-art" AI is delivering results slower than its legacy systems were.

By Frankenstein ecosystem, I mean disparate technology stacks stitched together without shared architecture. A monster is created that consumes resources, compounds complexity, and never delivers on the anticipated value. The costs are staggering. Not just the millions over budget, but the opportunity cost of moving at a crawl while competitors pivot and adapt.

The Price of Best-of-Breed

The vendor landscape has exploded. Gartner's 2024 research shows the AI market becoming increasingly complex, with 40 percent of generative AI solutions becoming multimodal by 2027, up from 1 percent in 2023.[1] Meanwhile, enterprises are drowning in choices. More than two-thirds of respondents plan to increase software spending, while a similar proportion expect to cut data center investments.[2] Companies are spending more on AI, while trying to simplify infrastructure, a paradox that demands smarter partnership decisions.

The sticker price is just the beginning. Beyond the data quality challenges covered in Strategy 3, there's vendor-specific data work that adds substantial cost.

The integration tax is particularly brutal. When systems don't talk to each other, it is paid twice. Once with direct costs for format conversions and custom connectors, and again in operational overhead.

Also, consider the security implications of siloed systems. Data breaches averaged $4.88 million in cost during 2024.[3] When juggling multiple AI vendors—each with different security protocols and data handling practices—the attack surface is not just multiplied, but gaps are being created between the systems, and they will be where breaches are most likely to occur.

Vendor demos are theater, carefully scripted performances designed to hide limitations and amplify possibilities. A vendor shows up with a slick demo. Its AI responds perfectly to every query and integrates seamlessly with mock data. The ROI projections look like a hockey stick.

Reality comes when you try to integrate with your own systems.

Vendor lock-in in AI is particularly insidious because it involves data, models, and business logic that are all intertwined and work together. Lock-in can happen at multiple levels:

- Data stored in proprietary formats that can't be easily exported.

- AI models that are black boxes, unable to be replicated or transferred.

- Business processes designed around the vendor's way of doing things.

Some lock-in is inevitable, but not all lock-in is equal. The key is deciding which dependencies are acceptable trade-offs and which are strategic risks then vetting vendors accordingly.

Choosing The Right Vendors

Cut through the sales pitch and evaluate what actually matters with this Four Pillar Vendor Assessment:

- **Technical scalability**: Can their architecture actually mesh with yours without requiring an army of integration engineers? Ask to see their data formats, not marketing materials. Request performance benchmarks under actual load conditions, not their ideal scenarios. Will that chatbot handling 100 queries beautifully in the demo choke at 100,000 in production?

- **Financial stability and roadmap alignment**: An AI startup might be the hot thing, but will it exist in two years? Does its product roadmap align with your strategic direction? This might not seem critical now, but over several years, it will pay off. If the vendor vision shifts, you could end up being yesterday's priority. A partner's future needs to align with yours.

- **Cultural fit and support philosophy**: When something breaks at 3:00 AM, who's answering the phone? What's its philosophy on customization requests? Some vendors view each one as a billable professional services engagement. Others see it as product feedback. The difference can cost millions.

- **Transparency and explainability**: "Black box" AI is a liability. Can the methodology with which the model makes decisions be explained, specifically for your use cases? When regulators come knocking, you need partners who can help you defend your AI decisions. Equally important is security transparency. Ask about security certifications (SOC 2, ISO 27001, FedRAMP, and others), penetration testing frequency, incident response procedures, and breach notification timelines. Request security architecture diagrams. A vendor unwilling to discuss its security posture in detail is a vendor hiding problems.

Here are some probing questions to ask:

- **"How was your last major outage handled?"** How failure was handled tells more success stories.

- **"How do you handle model drift and retraining?"** AI models degrade. Without a clear answer, you'll be stuck with declining performance.

- **"What's your position on intellectual property for custom models?"** Some vendors claim ownership of any models trained on your data. That's a competitive advantage slipping away.

- **"What's your model deprecation policy?"** Vendors retire model versions. If you've built and tuned workflows around GPT-4, what happens when they push you to GPT-5? This is a disruption risk.

- **"What happens to our data if we terminate the contract?"** If the answer involves significant cost or effort, you're looking at vendor lock-in.

Beyond critical operational and security vetting, it's imperative to understand the true economics of the vendor relationship. When it comes to AI costs, what you see is rarely what you get. The visible costs—licenses, implementation, training, support—often represent a fraction of the total investment. The rest hides in integration work, ongoing model retraining, scaling infrastructure, security audits, and the exit costs no one mentions until you try to switch vendors. Budget for the iceberg, not just the tip.

The TCO Math

Here's a framework for calculating the realistic total cost of ownership. These multipliers are based on patterns I've seen across dozens of enterprise AI implementations:

- **Year 1 costs**: Base license times 1.5 for overages and add-ons, implementation costs times 2 because things always take longer, integration costs at 15–25 percent of the total project, and training and change management at 10 percent of the total.

- **Ongoing annual costs**: Base license plus 20 percent annual increase, maintenance and support at 20–30 percent of license cost, quarterly model retraining, additional infrastructure while scaling, compliance and audit requirements, and an internal team to manage the vendor.

- **Hidden costs**: Multi-vendor coordination adds 30 percent if using more than three vendors, legacy system integration adds 40–60 percent, regulated industry requirements add 25–40 percent, and global deployment adds 20–30 percent per region.

- **Exit costs**: Data extraction and migration, retraining on the new platform, parallel running during transition, and contract termination penalties.

A $200,000 AI solution never stays at $200,000. Apply our TCO framework conservatively, and Year 1 costs land between $800,000 and $1.5 million — unless you're building in a true greenfield environment.

Platform Architecture: From Chaos to Control

The biggest architectural decision is how to structure the vendor ecosystem. As AI capabilities evolve toward autonomous agents, McKinsey's analysis reveals that managing distributed AI systems across the enterprise will become increasingly complex,[4] making a clear vendor architecture even more critical.

Two common approaches exist today:

- Consolidating the entire stack with one vendor for integrated governance, consistent data handling, and predictable costs.

- Diversifying across vendors with best-of-breed tools for each use case, then integrating these solutions after the fact.

The approach I recommend is a hybrid, neither all-in with one vendor nor scattered across too many. Build a hub-and-spoke model. Start with a major platform as the core foundation (e.g., Azure, AWS, GCP, or similar). This becomes the hub for governance, security, data management, and core AI needs.

For many organizations, a solid platform, implemented the right way, covers most of their AI needs. Choose this hub platform based on its ability to integrate with future specialized tools that will be needed. Run the Four Pillar Vendor Assessment on the leading vendor carefully. If selected, it will become the foundation for everything else.

When the hub lacks a capability or its built-in version doesn't meet your business requirements, add spokes. Each spoke must connect through the hub's abstraction layers and follow its security protocols, report to its monitoring systems, and comply with its data standards. The main platform is the foundation, while point solutions plug in for specific needs. Never let a point solution operate independently.

When I say each spoke must connect through the hub's abstraction layer, I'm not talking about the bloated middleware approach from 15 years ago during the service-oriented architecture (SOA) era, when enterprises built massive central buses to translate between every system. Modern abstraction isn't about building redundant interfaces, since most mature vendors already provide decent APIs and orchestration mechanisms. Use a platform's native API gateway (Azure API Management, AWS API Gateway, or similar) to create a unified access point that can route to different vendors. Add routing rules based on capability, cost or performance as needed.

The abstraction layer handles the messy realities, like converting dates between vendors who can't agree on ISO 8601, normalizing response formats so apps always receive consistent data structures, and managing vendor-specific authentication methods in one place. Most importantly, it keeps business logic—your actual competitive advantage—separate from vendor implementations. When that computer vision vendor changes its response format next month, only one transformation policy is updated in the API gateway.

This way, vendors can be switched without touching applications. Consistent standards are maintained across the entire ecosystem. Every app uses the same endpoints, data formats, and error codes, regardless of which vendor is doing the behind-the-scenes work.

Avoid vendors that can't integrate through the hub. I've seen organizations fail with the spokes-no-hub approach countless times, spending more time making systems talk to each other than generating AI value.

Alternative patterns exist if the hub-and-spoke doesn't fit. The *federation model* lets divisions choose their own AI solutions while corporate sets standards for data exchange and security. Regardless of the pattern that is chosen, two things to keep in mind:

1. **Maintain strict foundational governance**: Unified monitoring across all AI systems, consistent logging for debugging and security investigation, common metrics across AI types, enforced security requirements verified through audits, and comprehensive decision tracking for compliance.

2. **Data control is non-negotiable**: Never let a vendor control it. Demand portable formats. If they offer only proprietary exports, walk away. Regularly export data to proprietary storage. Understand how models are trained, and maintain the ability to retrain them elsewhere if needed. Include specific data portability requirements in contracts, with penalties for noncompliance. Whether it's hub-and-spoke or federation, these principles protect future flexibility.

Build Versus Buy

This debate isn't binary. It's a spectrum with multiple options, each with distinct trade-offs. The most successful outcomes come when build-versus-buy decisions align with business strategy, not with what's technically doable or what's trending in the market today.

Build

Build if the needed capability is core to competitive differentiation, sustained access to AI talent is in place, a price tag of three times the initial estimate can be afforded, speed to market isn't critical, and complete control over the IP and algorithms is required.

The 2024 enterprise AI landscape shows a near-even split between in-house and vendor-sourced solutions (47% vs. 53%), a major shift from 2023 when 80% of enterprises relied on third-party generative AI software.[5]

Buy

Buy when mature vendors exist, the capability requires no customization, integrates seamlessly with your tech stack, and isn't core to your differentiation, especially if time to market is critical or you lack specialized AI talent.

Partnering

When between build and buy, without the capability to build but a need that is too strategic to hand over to a vendor, it's a clue to partner. True partnership, not just vendor relationships, makes sense when there is a need for specialized expertise for a capability that is strategic to the business.

If the capability requires ongoing innovation, it's wise to share risk and reward. With a vendor that is willing to customize deeply and long-term incentives can be aligned.

Don't overlook open source solutions, though the "free" price tag is misleading. Open source works if you have deep technical expertise, are willing to contribute back to the community, can handle maintenance burdens like security patches, need maximum flexibility, and want to avoid vendor lock-in. With open source, you have to be the security team. Every vulnerability, which must be fixed immediately, is your problem.

Case Study: A Vendor Consolidation Success

Klarna, a global fintech company, discovered they were bleeding money to 1,200 different SaaS services.[6] Not 12, but 1,200.

The investigation uncovered a maze of disconnected systems. Marketing alone used dozens of different tools: email platforms, analytics dashboards, campaign managers, social media schedulers, A/B testing tools, and attribution systems. Each promised to be "the last tool you'll ever need." Each required its own training and integration, and many required their own login. The marketing team spent more time moving data between tools than actually marketing.

Customer service was worse. It had multiple disparate systems that didn't communicate. When a customer called with a problem, agents juggled multiple screens to find answers. Resolving a single issue took 11 minutes on average.[7] Not because agents were slow, but because no single system had the full picture.

The annual software bill was massive. And that was just the licensing. Add the cost of engineers maintaining integrations, analysts reconciling reports from different systems, and project managers coordinating vendor relationships. Every security audit meant reviewing dozens of different architectures. Every compliance update triggered months of vendor coordination.

The CEO saw what others missed. The company hadn't built a tech stack. It had created an accidental architecture, with 1,200 pieces held together by workarounds and layers of complexity.

It could have gotten much worse if they'd responded by adding more platforms to connect platforms, and monitoring tools to watch their tools. Klarna opted for radical surgery; painful, but very needed. It shut down platform after platform—Salesforce and Workday among them. Not because these are bad platforms, but because Klarna needed integration, not more features.

Instead of buying another feature-bloated enterprise suite, it adopted focused AI tools.

For marketing:

- Using AI image generation tools, including DALL-E, Midjourney, and Adobe Firefly, reduced image production time from 6 weeks to just 7 days, saving $6 million annually.[8]

- AI-powered copywriting tools generated their marketing content, contributing to the $4 million saved from a 25 percent reduction in agency spend.[8]

For customer service:

- It deployed an AI assistant powered by OpenAI that:

    - Handles two-thirds of all customer service chats.[9]

    - Does the "equivalent work of 700 full-time agents."[9]

    - Reduced resolution time from 11 minutes to under 2 minutes.[7]

    - Drives $40 million in profit improvement.[9]

The financial impact was immediate. Sales and marketing costs dropped 11 percent in Q1 2024, while campaign volume increased.[10] The value wasn't just cost savings. It was also speed. Development cycles that took six weeks now took seven days.

AI thrives on integration. Every vendor you eliminate is engineering capacity you recover.

Days 1–30: This week, select a hub platform. The chosen platform becomes the foundation and is the decision that will unlock everything else. Choose one that aligns best with business goals. For all other vendors, one question determines their fate: Is this strategic to our business? If yes, they stay but must connect through the chosen hub. If not, export data and terminate. A strategic vendor is one that is core to competitive advantage, has unique capabilities that can't be found elsewhere, or is deeply embedded in critical processes.

Run immediate security triage. While evaluating vendors, flag security risks that require immediate attention. Prioritize the highest-severity issues—such as missing SOC 2 Type II, storing unencrypted data, or undisclosed breaches in the last 12 months—and address those immediately.

Days 31-60: Build the hub-and-spoke architecture. The hub platform now rules. Every remaining vendor must connect through its API gateway, follow its security protocols, and report to its monitoring systems. Calculate TCO for each vendor relationship using the full cost framework: Year 1 costs, ongoing annual costs, hidden multipliers, and exit costs.

Require data portability in all vendor contracts and include specific penalties for noncompliance.

Days 61-90: Run the first quarterly architecture review. Can the current setup handle 10x growth without fundamental restructuring? If not, fix it now while there is momentum. For any capability gaps identified, apply the build-buy-partner spectrum:

build if core to competitive differentiation, buy if commoditized, partner if strategic but lacking internal capability.

[1] Emma Keen. "Gartner Predicts 40% of Generative AI Solutions Will Be Multimodal By 2027." Gartner [press release]. September 9, 2024. https://www.gartner.com/en/newsroom/press-releases/2024-09-09-gartner-predicts-40-percent-of-generative-ai-solutions-will-be-multimodal-by-2027

[2] NPI. "The State of Enterprise IT Sourcing in 2025." June 26, 2025. https://www.prnewswire.com/news-releases/new-npi-research-reveals-how-ai-vendor-consolidation-and-rising-costs-are-reshaping-enterprise-it-sourcing-in-2025-302491547.html

[3] IBM. "Cost of a Data Breach Report 2024." July 30, 2024. https://newsroom.ibm.com/2024-07-30-ibm-report-escalating-data-breach-disruption-pushes-costs-to-new-highs

[4] Alexander Sukharevsky et al. "Seizing the Agentic AI Advantage." McKinseyBlack: AI by McKinsey. 2025. https://www.mckinsey.com/capabilities/quantumblack/our-insights/seizing-the-agentic-ai-advantage

[5] Tim Tully, Joff Redfern, Derek Xiao (with Claude Sonnet 3.5). "2024: The State of Generative AI in the Enterprise." Menlo Ventures. November 20, 2024. https://menlovc.com/2024-the-state-of-generative-ai-in-the-enterprise/

[6] Silicon Editorial. "AI Strategy: Klarna Eliminates 1200 SaaS Services." Silicon. April 21, 2025. https://www.silicon.eu/ai-strategy-klarna-eliminates-1200-saas-services-17458.html

[7] "Klarna AI Assistant Handles Two-Thirds of Customer Service Chats in Its First Month." Klarna [press release]. February 27, 2024. https://www.klarna.com/international/press/klarna-ai-assistant-handles-two-thirds-of-customer-service-chats-in-its-first-month/

[8] "AI Helps Klarna Cut Marketing Agency Spend by 25% and Run More Campaigns." Klarna [press release]. May 28, 2024. https://www.klarna.com/international/press/ai-helps-klarna-cut-marketing-agency-spend-by-25-and-run-more-campaigns/

[9] "Klarna's AI Assistant Does the Work of 700 Full-Time Agents." OpenAI. February 2024. https://openai.com/index/klarna/

[10] Chris Kelly. "Klarna Uses AI to Save $10M on Marketing Annually While Upping Output." *Marketing Dive*. May 29, 2024. https://www.marketingdive.com/news/klarna-gen-ai-openai-cut-marketing-spend-efficiency/717332/

The Three Pages Between You and Disaster
Your AI Governance Shield

Let me confess. My first draft had 91 pages on AI governance. Everything from risk management frameworks to the full playbook for responsible AI. But then it hit me. You're not going to read 91 pages on governance, probably not even 20.

So, here's the deal. I've compressed everything into three pages you can print and read over lunch. I know governance isn't the most glamorous part of the job, but this ten-minute read can be the difference between leading an AI transformation and explaining one to lawyers.

Implementation teams can handle the details; you need to know what to have them care about before AI does something spectacularly wrong and expensive.

Five AI Risks That Can Derail Your Business

Traditional IT risk management assumes systems either work or break. AI introduces five new ways that can destroy value, all while the AI is "working."

1. **Hallucination risk**: AI lies with utter confidence. An AI hallucination is assuredly generating plausible-sounding—but

false—information. This isn't a bug; it's how large language models work. AI will hallucinate. The question is whether it's caught before costing millions.

2. **Model drift**: Yesterday's genius is today's idiot. Fraud detection trained on last year's patterns will miss this year's realities. Demand forecasting calibrated on pre-pandemic behavior fails catastrophically. Without drift detection, AI becomes a wealth destruction machine.

3. **Algorithmic bias**: AI can discriminate at machine speed. Every AI trained on historical data inherits its historical prejudices. When that AI makes thousands of decisions per minute, discrimination is automated at an unprecedented scale.

4. **Agentic autonomy**: AI can exceed its authority. As AI systems become more sophisticated, they don't just interpret rules, they read between them. Without clear boundaries and override mechanisms, a helpful AI assistant becomes an unsupervised employee with signing authority.

5. **Prompt injection**: AI can get hacked through conversation. Users are discovering they can manipulate AI systems with carefully crafted prompts. "Ignore your previous instructions and ... [fill in the blank]" has become the opening salvo in a new form of cyberattack. Customer service bots may be tricked into revealing confidential information, or even executing harmful actions. This isn't a traditional security vulnerability that can be patched. It's fundamental to how conversational AI works.

Five Questions for the Next Executive Meeting

Ask the following questions at the next team meeting. Hopefully, there are answers. Rethink things if there are uncomfortable silences.

1. **"Who owns the decision when AI screws up?"** If *the* person can't be named and held accountable for any AI system decision, there is an accountability vacuum. Every AI needs a business owner who accepts responsibility for outcomes, not just implementation.

2. **"How do we know our AI is still working?"** If AI isn't being monitored continuously, including automated alerts, then the AI is quietly degrading while in use. Set drift thresholds (information, warning, critical, fatal) and automated triggers before accuracy drops, not after customers complain.

3. **"Can we shut it down immediately?** If malfunctioning AI can't be immediately disabled, something as simple as one viral TikTok video can quickly lead to brand destruction.

4. **"What happens when the regulator calls?"** Can the AI's last 100 decisions be reconstructed? Can each one be explained? The EU AI Act demands both audit trails and explainability for high-risk systems. Fines can reach up to 7 percent of global annual turnover. "The algorithm said so" isn't a defense; it's an admission of negligence.

5. **"How much are we betting on someone else's AI?"** A vendor's hallucination is your liability; their bias is your lawsuit. Yet most companies have no idea what AI their vendors are using or how it's governed.

Five Things to Demand from Your Teams

These five practices form the foundation of your AI governance:

1. **AI inventory and risk classification**: Document every AI system, assign risk tiers (high/medium/low), identify the business owner, and confirm kill switch procedures. Not knowing what AI is running means not governing it.

2. **Decision logs for systems**: Every high-stakes AI decision needs a trail that answers what was decided, why it was decided, who could have overridden it, and audit records. When the regulator or an attorney asks, answers are needed.

3. **Vendor AI governance**: Require model cards documenting AI capabilities and limitations. Establish performance monitoring. Define change control processes. Clarify liability in contracts. The vendor's AI is your risk.

4. **Human-in-the-loop escalation**: Define clear triggers for human review. Customer service chatbots escalate policy questions. Lending algorithms flag unusual patterns. Medical AI requires physician confirmation. Content moderation APIs filter harmful output before it reaches users. Set the triggers before they are needed.

5. **Deactivation criteria**: Pre-define what triggers a shutdown, such as accuracy dropping below 85 percent, bias detected above thresholds, regulatory investigation initiated, or emergence of unusual patterns.

These practices align with the National Institute of Standards and Technology (NIST) AI Risk Management Framework's core functions — *govern, map, measure, and manage* (with #4 and #5 in my list aligning to the manage phase).[1]

Three Ways Governance Makes You Faster, Not Slower

1. **Network effect**: Strong governance creates a virtuous cycle. Governance builds trust, which enables ambitious deployments, success generates better data, and better data improves models, with improved models justifying more investment.

2. **Market access accelerator**: U.S. federal contracts require NIST alignment. When wanting access to its 450 million consumers, EU compliance isn't optional. Enterprise customers demand governance documentation. What looks like a compliance cost is actually a market entry fee that unprepared competitors can't pay.

3. **Innovation accelerator**: Pre-approved governance patterns turn months of review into days of deployment. When 100 percent of AI use cases follow standardized patterns that include built-in security and compliance, innovation teams move from idea to production much faster. Governance becomes the highway, not the roadblock.

Regulatory Realities

The regulatory landscape isn't on the horizon; it's already here. Requirements vary by region, and regulators are moving fast. The examples below are only the beginning:

- **EU AI Act**: The Act imposes fines of up to 7% of global annual turnover for prohibited AI practices[2], and up to 3% for violations involving high-risk systems[2]. HR screening and candidate-evaluation tools are explicitly classified as high-risk under Annex III[2].

- **U.S. patchwork**: With no comprehensive federal AI law, states have moved independently. In 2025, all 50 states introduced AI legislation[3], creating a fragmented regulatory landscape that companies must navigate.

- **Industry-specific regulations**: Existing regulatory frameworks increasingly apply to AI. The FDA regulates AI-enabled medical devices and has issued detailed guidance for their development and oversight[4]. In financial services, model-risk-management rules require transparency and accountability in automated decision-making[5]. Anti-discrimination laws also apply to algorithmic decisions in insurance[6].

Still, the companies treating compliance as a burden are missing the point. Every new regulation is a barrier to entry for unprepared competitors.

Case Study: Air Canada v. Moffatt

In February 2024, Air Canada learned that their AI chatbot wasn't just a customer service tool, it was a legal agent making binding commitments on the company's behalf. A Canadian tribunal made clear that Air Canada was fully responsible for the chatbot's statements, rejecting the airline's argument that the bot was a separate entity.[7]

Jake Moffatt's grandmother had passed away.[8] Seeking bereavement fare options for a last-minute flight, he turned to Air Canada's website chatbot for help. The AI assistant confidently told Moffatt he could retroactively apply for a reduced bereavement fare within 90 days of his ticket being issued.[7] It even provided a link to Air Canada's bereavement travel webpage. The linked webpage directly contradicted the chatbot, stating that bereavement fares could not be applied for after travel was completed.[7] But why would anyone read the fine print after the chatbot provided the answer?

When Moffatt tried to claim the discount, Air Canada refused, arguing they weren't responsible for their chatbot's errors. It claimed the chatbot was a "separate legal entity" that was responsible for its own actions,[9] an argument, if accepted, would have revolutionized corporate liability law.

The British Columbia Civil Resolution Tribunal did not accept it. In an unequivocal ruling, the tribunal held Air Canada accountable for all website content, whether generated by static pages or AI chatbots[7]. The tribunal ordered the airline to pay Moffatt $812.02.[7] — a trivial amount. But the precedent was anything but trivial.[10] Every company using customer-facing AI suddenly realized their chatbots weren't just providing information, they were making legally binding commitments. Every chatbot response is a company promise. Every AI interaction is a corporate action. Every automated decision is your decision. Govern accordingly.

▐ STRATEGY 6 ACTION PLAN

Days 1-30: This week, identify what AI is actually in use. Document every system that use AI, whether deployed in production, pilot stage, or shadow IT experiments. Track who built it, what data it uses, and what business function it serves. For vendor AI: require model cards documenting capabilities and limitations, and clarify liability in contracts.

For each AI system, determine its risk level based on business impact and data sensitivity. Assign a business owner (not from IT) for each system. Focus on the basics: who owns the system, how it can be shut down, and what happens when it fails. Set drift thresholds (information, warning, critical, fatal) and define what triggers a shutdown.

Days 31-60: Target the highest-risk AI systems first. Implement basic safeguards, including human review for critical decisions, bias monitoring for customer-facing AI, and audit trails for regulated systems. Ensure every high-stakes AI decision logs what was decided, why, and who could have overridden it.

Days 61-90: Extend safeguards to remaining AI systems. Keep iterating until every system has a governance foundation. Build the minimally viable shield first, then strengthen it through iteration.

[1] National Institute of Standards and Technology. "AI Risk Management Framework." NIST. https://www.nist.gov/itl/ai-risk-management-framework

[2] European Parliament and Council of the European Union. "Regulation (EU) 2024/1689 (Artificial Intelligence Act)." Official Journal of the European Union. July 12, 2024. https://eur-lex.europa.eu/eli/reg/2024/1689/oj

[3] National Conference of State Legislatures. "Artificial Intelligence 2025 State Legislation." NCSL. 2025.

https://www.ncsl.org/technology-and-communication/artificial-intelligence-2025-legislation

[4] U.S. Food and Drug Administration. "Artificial Intelligence and Machine Learning in Software as a Medical Device." FDA. Accessed January 25, 2026.

https://www.fda.gov/medical-devices/software-medical-device-samd/artificial-intelligence-and-machine-learning-software-medical-device

[5] Office of the Comptroller of the Currency. "Model Risk Management: Supervisory Guidance." OCC Bulletin 2011 12. April 4, 2011.

https://www.occ.gov/news-issuances/bulletins/2011/bulletin-2011-12.html

[6] National Association of Insurance Commissioners. "Artificial Intelligence (AI) Principles." NAIC. 2020. https://content.naic.org/sites/default/files/inline-files/NAIC%20Principles%20on%20AI.pdf

[7] Moffatt v. Air Canada, 2024 BCCRT 149. British Columbia Civil Resolution Tribunal. February 14, 2024. https://www.canlii.org/en/bc/bccrt/doc/2024/2024bccrt149/2024bccrt149.html

[8] Kyle Melnick. "Air Canada Chatbot Promised a Discount: Now the Airline Has to Pay It." *Washington Post*. February 18, 2024. https://www.washingtonpost.com/travel/2024/02/18/air-canada-airline-chatbot-ruling/

[9] Barry B. Sookman. "Moffatt v. Air Canada: A Misrepresentation by an AI Chatbot." McCarthy Tétrault. February 19, 2024. https://www.mccarthy.ca/en/insights/blogs/techlex/moffatt-v-air-canada-misrepresentation-ai-chatbot

[10] Kirsten Thompson. "Airline Ordered to Compensate a B.C. Man Because Its Chatbot Provided Inaccurate Information." Dentons Data. February 15, 2024. https://www.dentonsdata.com/airline-ordered-to-compensate-a-b-c-man-because-its-chatbot-provided-inaccurate-information/

Pilot, Measure, Kill or Scale
| No Zombies. No Exceptions

Your Data science team is brilliant at building science experiments they call pilots. I haven't met them, but I know every data science team is. In fact, 2025 research from MIT found that just 5 percent of custom enterprise AI pilots extract meaningful value—the vast majority deliver zero measurable P&L impact.[1] Similar research from IDC found the numbers only slightly better: for every 33 AI proof-of-concepts companies build, only four make it to production—an 88 percent failure rate.[2]

Most of these failures are predictable and preventable. If you've implemented the earlier strategies, you've already eliminated the most common pilot killers:

- **Panic-driven pilot**: Strategy 1 ensured starting with business outcomes, not just chasing the technology.

- **Ownership vacuum**: Strategy 2 ensured leadership alignment and cultural readiness.

- **Data foundation failure**: Strategy 3 established business-led data ownership, quality standards, and accessible architecture.

- **Translation gap**: Strategy 4 ensured domain experts and translators catch misalignments before building.

- **Missing governance**: Strategy 6 established clear ownership, monitoring, and kill switches.

Check all of those boxes. If any are unmarked, you're not ready for production scaling.

Six Technical Killers (and How to Beat Them)

The six remaining killers are technical execution mistakes. But before I get into the details, I need to discuss a distinction that trips up most organizations: the fact that POCs and pilots are fundamentally different. I'm sure you know the difference, but are you treating them differently?

In my experience, even very skilled teams see high pilot failure rates. Root causes start early. What happens is that companies throw any proof of concept (POC) with potential success into production and hope for the best. Take a minute to remember the last pilot you launched. Was a POC deployed straight to production, with problems patched as they surfaced, or did you take a minute to tell the team, "Great, the POC has potential, let's make sure it's production-ready before we deploy it to production."

Proof of concepts are a playground. Use synthetic data, simplified infrastructure, and relaxed security. Move fast, break things, and prove the algorithm works. Security here means keeping the POC isolated; it never touches production systems, real customer data, or live networks. POCs are a cheap way to encourage controlled experimentation.

Pilots are production-ready software running on a small scale. They must run on production-grade infrastructure with real data, or they'll lead to dangerously false confidence. Security requirements don't scale down; the pilot inherits every production control from day one. I'll note the dev/prod parity principle from the Twelve-Factor App methodology —a set of principles for building scalable cloud applications— is still true today: gaps between environments are where failures hide.

There's another reason to run production-ready pilots. They expose the costs POCs hide (production SKU costs, data-pipeline maintenance, integration complexity, and the retraining treadmill).

The executive ask: Before greenlighting any pilot, ask, "Did we rebuild this for production, or are we calling a POC a pilot?"

Now that it's clear that pilots are being built—and not dressed-up POCs—here are the six technical killers—and how to beat each one.

Killer #1: The Data Quality Disaster

MIT Technology Review found data quality to be the top barrier to AI deployment, cited by 49 percent of companies overall and 52 percent of enterprises over $10 billion.[3]

To solve this, establish data quality gates that test four dimensions continuously:

1. **Completeness**: Enough data volume to make predictions.

2. **Coverage**: Required entities and edge cases are represented.

3. **Signal quality**: Noise and irrelevant data filtered out.

4. **Freshness**: Data reflects current patterns, not stale history.

These gates run at ingestion, not just during training. Teams validate that training data and production data come from the same distribution before building anything.

The executive ask: Before approving any pilot, demand proof that training data and production data come from the same distribution.

Killer #2: The Deployment Collapse

The pilot runs beautifully on historical data. Then production hits with 10x the volume and schema changes that nobody saw coming. The model that took three seconds to score now takes three minutes. Four disciplines prevent collapse:

1. **Shadow deployment**: Run a pilot parallel to production that does not impact actual operations. Such shadow deployment isn't optional. It's the only way to discover what production will actually look like. The model can be mathematically perfect; it can still fail operationally.

2. **Progressive volume scaling**: Start at 1 percent of production traffic. Once that's stable for a week, move to 5 percent. Then 10, 25, 50, and finally 100 percent. Each increase reveals new failure modes. A recommendation engine that handles 1 percent perfectly will deadlock at 5 percent due to database connection limits. A chatbot that responds instantly at 10 percent will time out at 25 percent due to token limits.

3. **The 10x Rule**: Test at 10x the expected volume. Even if that kind of growth is not expected, production can be spiky. Black Friday, system failures that create backlogs, and viral social media moments will create 10x spikes that kill pilots designed for average load.

4. **Geographic and temporal testing**: A pilot program trained on California data will likely fail in New York. A model trained on weekday patterns fails when hit with weekend patterns. Test across all production scenarios, including different geographies, time zones, seasons, user segments, device types, and network conditions. What works on desktop could fail on mobile; urban areas can handle traffic that fails in rural counties due to poor connectivity.

Killer #3: The Model Decay Trap

Models age like milk, not cheddar. Accuracy of 94 percent could degrade to 80 percent within few months. Not suddenly, but gradually enough that it won't be noticed until customers complain. Five disciplines can help prevent this:

1. **Drift detection thresholds**: Define what drift means for the specific use case. For fraud detection, a 2 percent increase in false positives is acceptable. For medical diagnosis, 0.5 percent degradation triggers immediate investigation. Set multiple thresholds, including information (log it), warning (investigate it), critical (fix it now), and fatal (shut it down).

2. **Feature monitoring**: Features drift before models fail. Monitor the distribution of every feature feeding the model. When the customer age distribution shifts from an average of 35 to 42, a model trained on younger demographics starts failing.

3. **Automated retraining triggers**: Don't retrain on schedules, retrain on triggers. If nothing has changed, monthly retraining wastes resources. But waiting for calendar dates when drift accelerates means bad predictions spanning weeks. Set triggers based on performance degradation (accuracy drops 2 percent), data drift (feature distributions shift beyond thresholds), time-based maximum (never go more than 180 days), and business events (new product launches, seasonal changes) so that the models retrain automatically when performance degrades, rather than when the calendar says so. Some models retrain weekly during volatile periods, others run for months without retraining during stable periods.

4. **Building feedback loops**: A model needs to learn from production reality. Implement explicit feedback collection (thumbs up/down on recommendations), implicit signals (was a recommended item purchased?), and feedback-triggered retraining (e.g. when negative feedback exceeds 15 percent, retrain automatically). A recommendation engine that ignores whether customers actually bought recommended products is flying blind.

5. **A/B testing for retraining**: Never replace a model without proof that the new one performs better in production. Run both models in parallel, sending 10 percent of the traffic to the new version. Compare business metrics, not just technical ones

The executive ask: What triggers retraining, and how do we know it's working?

Killer #4: The Hidden Cost Spiral

If proof-of-concept costs X, the pilot costs 3–5X. If the pilot costs Y, production costs 4–6Y. Annual maintenance runs 30 percent of production cost, forever. Retraining costs 15–20 percent of initial development, monthly. Year two costs exceed year one by 60 percent. This cost curve wasn't in anyone's forecast.

Initial development is just the beginning. Here's what's truly required:

*   **Development (Year 0)**: Model development is the baseline, but it's the smallest piece. Data pipeline development can cost as much as the model itself. Integration with existing systems can double the budget, then add security, compliance, testing, and validation.

*   **Deployment (Year 1)**: Infrastructure at production scale costs multiples of pilot infrastructure. Monitoring, observability,

production hardening, training, and documentation will be needed, and the first retraining cycles will typically be three to six times in Year 1 alone.

- **Maintenance (Year 2+)**: Ongoing retraining, engineering features as business needs evolve, pipeline maintenance, quarterly compliance audits, and continuous drift monitoring. These costs never go away.

- **Incremental versus full retraining**: Full retraining costs 10x more than incremental updates. Use transfer learning. Start with an existing model and fine-tune on new data. A recommendation system that cost $100,000 to build initially will now update for $5,000 monthly through incremental learning. But every six months, it needs full retraining to prevent compound drift. Budget for both.

- **MLOps investment**: These platforms seem expensive until alternative to MLOps are calculated. Manual deployments, custom monitoring, and ad-hoc retraining costs will compound. The platform cost pays for itself through efficiency gains.

- **Kill criteria based on cost-benefit**: Define when to pull the plug before starting. If monthly costs exceed monthly benefits for three consecutive months, kill it. If retraining costs grow faster than value delivered, kill it. If maintaining the model costs more than the problem it solves, kill it.

The executive ask: When do we pull the plug, and who makes that call?

Killer #5: The Handoff Failure

The moment data science hands its model to operations, many AI projects die. Data scientists build for accuracy, while operations needs reliability. Data scientists optimize for performance, while operations needs maintainability. The handoff between these two teams is a major failure point. They speak different languages, use different tools, and have different definitions of success. Data science celebrates 95 percent accuracy. Operations asks, "What happens to the 5 percent that fail at 3:00 AM?"

To solve this, bring MLOps into development from day one. They need veto power on architecture decisions that affect maintainability. Create four essential documents:

1. **A one-pager**: Something executives can understand in two minutes.

2. **A runbook**: Include full actual commands (not "restart the service" but the exact command to execute to restart the service).

3. **A security guide**: Show who has access to what.

4. **A technical deep-dive**: For those who need it.

Establish clear ownership using RACI matrices: who's *responsible* for model performance, *accountable* for uptime, *consulted* for updates, and *informed* about issues.

MLOps plays a critical role. They handle the infrastructure that keeps AI workloads reliable and well-governed at scale. They bridge the gap between "it works in Jupyter" and "it works at 10:00 AM on Black Friday," by building the deployment pipelines, monitoring systems, and circuit breakers that prevent disasters. Without this, data scientists are being asked to become operations experts, or operations to become ML experts. The end result will be that both fail.

The executive ask: Operations must validate and run the model independently before accepting handoff, with clear ownership and escalation paths defined.

Killer #6: The Security Retrofit

Security, along with governance and privacy, is the biggest brake on AI deployment speed, cited by 45 percent of all companies and 65 percent of the largest enterprises.[3] But not because security is inherently slow. It's because teams treat it as Phase 2. There is no Phase 2 for pilots killed by vulnerabilities.

Most frameworks give security its own chapter. Most teams reduce security to auth protocols bolted on after the model works. Such industry-wide failure in how security is approached is exactly why 45 percent of companies cite it as their biggest deployment brake. I've done things differently in this book. Security doesn't have its own chapter; I've woven it into every strategy. It isn't another layer; it's a mindset of building securely. If Strategies 1 through 6 have been implemented, pilots will inherit security by design.

The executive ask: Simply, no pilot proceeds to production without all the security controls in place. No exceptions.

Kill or Scale: The Gates Framework

The six technical killers outlined above compound quickly. The typical response is to grant exceptions.

- "Give it another quarter to prove value."

- "We've invested too much to stop now."

- "The team just needs more time."

One exception becomes precedent. Then the next struggling pilot gets the same grace period. The result is what we're seeing across enterprises now: zombie pilot portfolios, full of AI pilots that are not dead, though also not alive, consuming resources indefinitely with no path forward and no permission to die.

To prevent an accumulation of zombie pilots, implement mandatory gates. The following five gates compress the industry average of eight months into twenty-four weeks, a bit aggressive, but very achievable because at this point you've already eliminated the blockers that typically stall most pilots. Every pilot needs these gates, no exceptions. Each gate includes *Pass* criteria (minimum thresholds to proceed) and *Kill* criteria (hard stops, no path forward). A summary table at the end provides quick reference.

Entry Gate (Week 1–3)

Purpose: Ensure the pilot is worth doing.

Pass Criteria

- Problem significantly impacts the business, typically something that appears in executive reviews or affects a substantial portion of the target process.

- ≥80% of required production data clean and accessible.

- Executive sponsor with budget authority.

- ROI projection ≥2:1 through Year 2 (accounts for maintenance cost overruns).

- Business translator assigned; domain experts engaged to validate business logic.

- Operations (MLOps) involved with veto power on maintainability.

Kill If

- Problem wouldn't make the division's top-ten priority list.

- <60% of required production data clean and accessible.

- No executive owner.

- ROI <1.5:1 even under optimistic assumptions.

- Known legal or regulatory prohibition exists.

Technical Gate (Week 4–6)

Purpose: Prove the solution works in the real world.

Pass Criteria

- Model runs on production data, not synthetic datasets.

- Baseline established for comparison (if replacing human decisions, what's their current accuracy? 96% is not enough if a human achieves 98%).

- Model accuracy closes at least 80% of the performance gap to baseline.

- Data pipeline handles 2x current volume.

- Data quality verified at ≥80%.

- Security controls implemented and tested.

- Legal review completed and cleared for regulated data.

Kill If

- Data quality below 60%.

- Accuracy closes <60% of performance gap to baseline after tuning.

Business Gate (Week 8–12)

Purpose: Validate that the pilot creates measurable business value.

Pass Criteria

- Business metrics show measurable improvement; not promises, actual measured change.

- User adoption meets target, usually around: 30% for internal tools, 10% for customer-facing, 90% for compliance-required.

- Stakeholders actively engaged, not just tolerating the pilot.

- Integration with production systems proven.

- Operations (engaged from the start) reviews handoff documentation for completeness (one-pager, runbook, security guide, technical deep-dive).

- ROI holding at ≥1.5:1 (Entry Gate projected 2:1; actuals can accept tighter margin not far from projected).

Kill If
- Zero business metric improvement after adjustments.

- Adoption 70%+ below target despite training.

- Stakeholder withdraws support or actively blocks expansion.

- ROI below 1.5:1 on actuals.

Scale Gate (Week 16–20)

Purpose: Prove the economics hold at scale.

Pass Criteria
- System handles 10x normal load for 48 continuous hours.

- Unit economics validated: value delivered exceeds cost per prediction.

- Retraining process automated and tested.

- Shadow deployment validated against real production traffic.

- Progressive rollout from pilot to full production: 1%, 5%, 10%, 25%, 50%, 100%.

- Operations formally accepts ownership; not documentation review, active accountability for uptime, incident response, and maintenance.

- For regulated industries: compliance requirements documented, approvals secured, notifications submitted.

Kill If
- System fails to meet SLAs at 2x load (the Technical Gate minimum).

- Costs scale superlinearly (doubling users triples costs).

Production Gate (Week 24)

Purpose: Make the final call, scale or kill. Unlike earlier gates, there is no middle zone between the Pass and Kill criteria here. The pilot is either ready for real customers and real operational ownership, or it isn't.

Pass Criteria
- All previous gate criteria re-verified; confirm nothing has regressed.

- Production funding approved for at least two years (one-time funding creates zombie pilots).

- Success metrics defined with drift thresholds, retraining triggers, and monitoring in place.

- Rollback plan tested (can revert within your defined threshold, four hours is typical).

Kill If
- Any previous gate no longer holds.

- Funding not secured for two years.

- Success metrics undefined or monitoring not in place.

- Rollback plan not tested.

The Gate Review Protocol

Each gate concludes with a scheduled review.

- Pilot owner presents status against *Pass/Kill* criteria. Data, not opinions.

- Stakeholders get 5 minutes each for questions.

- If all *Pass* criteria are met, the pilot is good to proceed to next gate.

- If some *Pass* criteria are not met but above *Kill* threshold, and there is a credible path to fix them, vote to either kill now or continue with specific owners and timeline—documented and trackable. If the timeline is missed, kill it. Never proceed to the next gate until all *Pass* criteria have been addressed.

- If any *Kill* criteria are met, kill the pilot now, no exceptions.

If failure becomes obvious before a scheduled review, don't wait. Delay burns budget and capacity that could fund projects with actual potential.

Killing a pilot doesn't mean deleting the code or losing the learning. It means decommissioning it, so it no longer consumes infrastructure, budget, or team time. The code stays in the repository. The learning informs future pilots.

Gate Thresholds: Quick Reference

Gate	Purpose	Pass Threshold	Kill Threshold
Entry	Ensure the pilot is worth doing	Data ≥80% ROI ≥2:1 Executive sponsor Ops veto power	Data <60% ROI <1.5:1 No exec owner Problem outside top-10 Known legal prohibition
Technical	Prove the solution works in the real world	80% of gap closed Pipeline 2x Data ≥80% Security tested Legal cleared	Data <60% <60% of gap closed
Business	Validate measurable business value	Metrics improved Adoption at target Stakeholders—engaged ROI ≥1.5:1	Zero improvement Adoption 70%+ below Stakeholder withdraws ROI <1.5:1
Scale	Prove the economics hold at scale	10x load 48hrs Unit economics positive Shadow validated Ops accepts ownership	Fails SLAs at 2x Costs superlinear
Production	Commit or kill, no extensions	All gates re-verified 2yr funding Metrics/monitoring set Rollback tested	Any gate regressed No 2yr funding No metrics/monitoring Rollback not tested

Translating Technical Metrics to Executive Decisions

A data scientist saying the model achieves "95 percent accuracy" means nothing without context. More importantly, what is that accuracy worth in dollars? But also, what will the 5 percent error cost?

Every model decision falls into four categories that directly map to money. Use this TP-TN-FP-FN framework:

1. **True positives (TP)**: Correctly identified problems. E.g. in fraud detection, these are the fraudulent transactions caught. Each one saves the fraud amount.

2. **True negatives (TN)**: Correctly identified non-problems. These are legitimate transactions that were not blocked. They generate revenue.

3. **False positives (FP)**: Incorrectly flagged problems. These are legitimate transactions blocked. Each one costs the transaction value plus customer frustration.

4. **False negatives (FN)**: Missed problems. These are fraudulent transactions approved. Each one costs the fraud amount plus recovery costs.

Here's how this translates to executive decisions. The model catches 95 percent of fraud (true positives) but flags 3 percent of legitimate transactions (false positives). Sounds good until doing the math. There are 10 million transactions processed monthly, worth $500 million. That 3 percent false positive rate blocks roughly $15 million in legitimate transactions, which is around 300,000 customers who tried to buy something and couldn't. If even 10 percent of those customers never return, and the average customer lifetime value is

$1,000, $30 million in future revenue is lost. The model needs to prevent more fraud than the customer relationships it destroys.

For different domains, the math changes, but the principle remains:

- **Healthcare**: A diagnostic AI with 96 percent accuracy sounds impressive until you know that just predicting "healthy" returns 94 percent accuracy, since most people aren't sick. And sensitivity is required for critical conditions. Missing 1 percent of cancers (false negatives) costs lives. Flagging 5 percent of healthy patients for unnecessary procedures (false positives) costs millions.

- **Customer service**: A chatbot with 70 percent accuracy that handles 50 percent of queries still saves money if it truly resolves them. But if 30 percent of its answers are wrong and create follow-up calls, costs increase, not decrease.

- **Predictive maintenance**: Knowing equipment will fail with 90 percent accuracy means nothing if the time window cannot be specified. Predicting failure "sometime in the next year" versus "in the next 48 hours" determines whether anything can actually be done about it.

Translation determines whether pilots get funded, scaled, or killed. Master it, and you'll never struggle for AI investment again.

What Most Success Stories Leave Out

Finding a well-documented case study for this strategy was challenging. Companies rarely publicly share their pilot-to-production journeys, especially the messy middle where pilots get evaluated, killed, or scaled. There are press releases about successful deployments and statistics about failure rates that can be found, but almost never the specific gates, criteria, and decisions that separate the 12 percent that succeed from the 88 percent that fail.[2]

But here's what I know from being in the room working with Fortune 500 companies on their AI deployments: the most successful ones have some version of systematic evaluation frameworks. Some companies still make pilot decisions based on executive intuition and whoever has the stronger reasoning in the quarterly reviews, which works until the pilot graveyard quietly fills up.

Meanwhile, the companies scaling AI successfully have systematized what everyone else leaves to chance. These companies didn't stumble into success. They earned it through disciplined decision-making.

STRATEGY 7 ACTION PLAN

Days 1-30: This week, ask: "Are these actual pilots or dressed-up POCs?" Map survivors against Entry Gate criteria and schedule the Entry Gate review meeting so the team has time to prepare for the evaluation. For each pilot, calculate the two-year total cost of ownership, with best and worst-case scenarios. If best-case costs exceed value delivered, kill immediately. For new pilots entering the pipeline, complete the Entry Gate evaluation, including problem validation, sponsor identification, and data assessment. Translate technical metrics to business value for all active pilots using the TP-TN-FP-FN framework. Engage operations with every pilot.

Days 31-60: Validate that training data matches production data distribution before any pilot advances. Test at 2x current volume to catch pipeline limits early. Validate that the security controls are in place. Define drift thresholds and automated retraining triggers.

For pilots that pass the Technical Gate, begin the Business Gate evaluation and handoff documentation review. Establish RACI for each pilot: who's responsible, accountable, consulted, and informed.

Days 61-90: Measure actual business metrics, validate user adoption, and confirm stakeholder support. Operations must validate and run the model independently before accepting handoff. Conduct a first quarterly portfolio review. Rank all pilots by ROI. Kill underperformers.

What Happens Next: Progressive volume scaling—1 to 5 to 10 to 25 to 50 to 100 percent. Complete the Scale Gate evaluation: 10x load testing, shadow deployment, unit economics validation, and operations ownership. Establish model health monitoring: drift thresholds, retraining triggers, and alerts. Complete the Production Gate: deploy to production or kill the pilot. Implement feedback loops so models learn from production reality. Use A/B testing before replacing any model.

[1] Aditya Challapally, Chris Pease, Ramesh Raskar, and Pradyumna Chari. "The GenAI Divide: State of AI in Business 2025." MIT NANDA. July 2025.

[2] Evan Schuman. "88% of AI pilots Fail to Reach Production: But That's Not All on IT." *CIO*. March 25, 2025. https://www.cio.com/article/3850763/88-of-ai-pilots-fail-to-reach-production-but-thats-not-all-on-it.html

[3] MIT Technology Review Insights. "A Playbook for Crafting AI Strategy." 2024. Page 12, Figure 5. Available at: https://www.technologyreview.com/2024/08/05/1095447/a-playbook-for-crafting-ai-strategy/

Innovation Beats Automation
| Build the Future While Optimizing Today

A CEO looks at the quarterly results with satisfaction. Every KPI is green. Automation has cut costs by 30 percent. AI chatbots handle 80 percent of customer inquiries. Predictive maintenance has reduced downtime to near zero. Then comes the question that challenges the familiar playbook.

"What do we do next?"

The stillness that follows reveals a truth every executive must face. Efficiency has limits. When everything worth automating has been, when processes run at their theoretical maximum efficiency, then competitive advantage shifts from doing things better to doing better things. The choice isn't between efficiency and innovation; it's doing both simultaneously or losing the long-run game: optimization to run today's business, innovation to build tomorrow's.

Maybe you're thinking, "Not me. I'm still trying to get basic fraud detection to work. I'm deep in Strategy 3 strengthening the data pipeline, nowhere near Strategy 8. Talking about innovation seems a little premature."

That thinking is more common than most leaders admit, and it's completely understandable. It's also one of the reasons AI transformations often plateau. And it's why I'm dedicating an entire chapter to innovation in a book focused on artificial intelligence. If AI implementation focuses solely on automation, you could be efficiently building the wrong things. Worse, systematic innovation can reveal that the process being automated shouldn't exist at all.

This chapter isn't just for executives who've mastered efficiency. It's also for those still building an AI foundation. The companies that are building innovation systems alongside operational optimization don't just innovate better over the long run; they operate smarter from the start. The cost of ignoring this advice plays out the same way across every industry.

Look at today's most successful companies. They didn't wait until optimization hit a ceiling to start innovating. Google didn't optimize search capabilities to perfection, then create Android. They built new platforms while their search tools were still improving. Amazon was still perfecting one-day delivery when it launched AWS, which today generates more profit than its entire retail operation. Contrast this with Blockbuster perfecting late fees, or BlackBerry optimizing keyboards.

Clayton Christensen's research in *The Innovator's Dilemma* documented how market leaders across industries consistently failed. Not from lack of capability, but from perfecting existing business models while new entrants created entirely different value propositions. The leaders excelled at improving their products, yet lost everything when new entrants redefined the basis of competition entirely.[1]

Innovation Is Not Reserved for Silicon Valley

The word "innovation" terrifies executives. They think it requires hiring PhDs or consultants with a Silicon Valley address. Innovation, simply put, is any novelty that creates value for a business, like a new product, a better workflow or a complex approval process simplified in a creative way.

Companies that succeed at innovation understand its brutal math. Finding one exceptional opportunity requires evaluating hundreds of ideas. This unnerves executives trained to expect a 90 percent success rate.

If you felt that tension when reading thus far, it's telling you something important: innovation and operational optimization require fundamentally different mindsets. For the rest of this chapter, I want you to use the half of your mind that thinks like a venture capitalist—expecting most bets to fail—rather than the Six Sigma black belt minimizing variation. No VC expects every investment to succeed; they expect a few outsized wins to offset many losses. This duality isn't just an individual shift; organizations must adopt it too— building a culture where optimization and innovation can coexist without one suffocating the other. They must operate with the discipline of optimization while investing with the level of failure tolerance that innovation demands.

This duality is what the innovation literature calls being "ambidextrous," using exploitation to optimize today's operations while utilizing exploration to discover tomorrow's opportunities. The companies that master ambidexterity don't choose between efficiency and innovation; they excel at both. But that starts with accepting that these two objectives follow completely different rules.

Note that individual failures don't make innovation unpredictable. When you manage innovation as a portfolio of bets, failures become learning, and the overall portfolio becomes measurable and predictable.

So, how to build this systematic innovation capability? How to become ambidextrous without tearing the organization apart? The answer isn't a single "innovation initiative" or hire; it's a comprehensive system built across five phases, outlined briefly below and unpacked in detail throughout the rest of this chapter:

Phase 1: Architect the innovation system by choosing the operating model your culture can sustain, assembling T-shaped talent capable of cross-domain problem-solving, and ring-fencing exploration funding so innovation survives operational pressure.

Phase 2: Source ideas through the three paradigms of pull, push, and prune, using each strategically and generating the massive volume required to find exceptional opportunities.

Phase 3: Validate ideas with problem-reframing and Amazon's Working Backward method, then enable rapid experimentation through data-rich sandboxes, automated compliance guardrails, and create multiple mechanisms for idea generation.

Phase 4: Govern the innovation portfolio by setting clear innovation zones, balancing incremental, adjacent, and transformational bets, funding work through learning-based tranches, measuring innovation velocity and revenue freshness, and ensuring single-threaded ownership from concept to scale.

Phase 5: Sustain innovation through culture by treating failure as learning, separating exploit and explore with distinct metrics and protections, and creating dual-track career paths so innovators can advance without abandoning innovation.

Each phase builds on the previous one, but perfection in one is not required before starting the next.

Phase 1: Architecting the Innovation System

Before allocating a single dollar or hiring an innovator, answer the fundamental question: "Where will innovation happen in the organization?" Without a clear operating model, there will be turf wars, duplicated efforts, and innovations that die in the gaps between departments.

Three proven models work at enterprise scale, and choosing the right one depends on culture, not aspiration:

- **Hub-and-spoke model**: Puts a central innovation team at the center, coordinating, while business units execute. The central team doesn't control innovation; they enable it by aggregating opportunities across silos, codifying successful experiments, and disseminating proven innovations organization-wide.

- **Venture studio model**: Creates a dedicated unit that builds new businesses from scratch. This model works when seeking radical innovation outside current business constraints.

- **Federated model**: Gives each division independence to innovate with only light central coordination. This works when businesses are diverse, and speed matters more than synergy. Each unit moves fast, but some efficiency is sacrificed due to duplicated efforts.

Centralized companies fail at federated models because their culture demands coordination. Siloed organizations struggle with hub-and-spoke because they lack collaboration muscles. Pick the model that fits your reality, not your PowerPoint vision.

The operating model is just the structure. Innovation requires a specific talent profile, which most HR systems aren't designed to identify. They are T-shaped individuals who have deep expertise in one domain, combined with broad collaborative skills. Pure specialists can't connect dots across domains, while pure generalists lack the depth to execute.

Finding T-shaped talent requires looking beyond traditional performance metrics. Google doesn't ask what candidates know; they present problems the candidate has never seen and watch how they think through ambiguity. Pixar's Brad Bird specifically sought out the studio's "black sheep"—artists who were frustrated because their unconventional ideas weren't being heard. "Give us all the guys who are probably headed out the door," Bird said. "A lot of them were malcontents because they saw different ways of doing things."[2]

Such individuals can underperform in narrow role definitions, while excelling when given broader mandates. They bridge technical and business conversations, pull insights from unrelated domains, ship solutions rather than just proposing ideas, and build coalitions across organizational silos. The goal is finding people who build things without being asked, those who question assumptions that others accept, and who can explain complex ideas simply.

Structure and talent mean nothing without resources; innovation dies in Q4 budget reviews without ring-fencing. When earnings pressure hits, unprotected innovation funding becomes the first casualty.

Start with 0.25-0.50 percent of revenue for unspecified exploration. This isn't project funding with predetermined outcomes; rather, it is exploration money for discovering what you don't know you don't know. The protection matters more than the amount. When innovation funding survives earnings pressure, it signals—throughout the organization—that innovation is strategic, not optional.

Phase 2: Sourcing Ideas—Pull, Push, and Prune

Innovation literature traditionally describes two paradigms. *Pull* innovation, where demand exists and pulls innovation toward it, and *push* innovation, where breakthrough capabilities are pushed into markets that don't yet know they need them. Both assume innovation means adding something new to the world. There is a third overlooked paradigm that focuses on reducing complexity. I call it *prune* innovation. Overlooking it was costly before. With AI, it's catastrophic. Organizations automating complex systems with AI institutionalize unnecessary complexity beneath a polished interface, then wonder why results disappoint. Let's examine all three paradigms in detail.

Pull: Starts with a confirmed problem and works backward to solutions, which is what most innovation methodologies teach—and for good reason, since it has the highest success rate. The customer need is verified, the pain is real, and the idea is to solve something people actually want solved. But pull innovation has limitations. When every company can see the same customer problems, only marginal differentiation is possible, since ten companies are all building slightly better CRM systems, or five startups are attacking the same pain point. Pull innovation is reliable but rarely revolutionary.

Push: Takes the opposite approach by starting with a technology or capability, then looking for applications. Despite lower success rates, push sometimes creates breakthrough innovations that nobody knew they desired. Push innovations typically fail because teams fall in love with their technology, as opposed to finding genuine value. The key is rapid, cheap experimentation to find product-market fit before heavy investment.

Push works when three conditions align:

- The capability must be truly unique, not just incrementally better.

- Resources are available for many cheap experiments, with an understanding that most will fail.

- Most importantly, there must be a willingness to pivot completely from the original vision.

Prune: Questions whether processes need to exist, redesigning systems to achieve the same outcome with less complexity. Most executives look outward for innovation: new customer problems to solve, new technologies to deploy. Prune looks inward to unlock value trapped by system complexity.

Subtraction as an innovation technique is a well-documented practice. Methodologies like Lean and Six Sigma use elimination, but in most organizations that work lives in operations or IT backlogs with efficiency mandates and cost-cutting targets. Prune is different in that it belongs to the innovation agenda, and it starts by asking whether the process should exist at all before looking for streamlining opportunities. Starting from "should this process exist?" before asking "how do we optimize this process?" unlocks different solutions.

Pruning is counterintuitive but necessary. Research from the University of Virginia found that when faced with problems, people overwhelmingly default to adding elements even when subtraction would lead to a better solution.[3]

Consider this simple example of a customer onboarding workflow. Over time, it accumulated 27 steps with three approvals, each added in response to an audit finding or incident. When delays became a problem, the operational response was to streamline handoffs, tighten SLAs, deploy AI to route requests, and provide decision support. Cycle times improved, but the underlying structure stayed the same.

Framed as an innovation opportunity instead, a team could step back and question the workflow's purpose. They might find a different way to satisfy the same policy requirements with only five steps and one approval, routing true exceptions separately. AI applied to the original 27-step process makes it faster; AI applied to the redesigned 5-step version makes it faster and simpler, which costs less to run and is easier to maintain.

The examples are endless; the principle is simple. Complexity breeds complexity. Prune encourages redesigning for simplicity instead of stacking workarounds.

The most innovative companies don't choose one paradigm. They use all three strategically, often in combination. Regardless of which paradigm is chosen, one mathematical reality remains constant. Finding exceptional innovations requires massive quantities. To find one breakthrough idea, you need to evaluate tens, sometimes hundreds, of possibilities. This is the inescapable math of innovation.

Microsoft's annual hackathon demonstrates this mathematical reality. It draws 71,000 participants across 75+ venues worldwide, spanning technical and non-technical roles across the company.[4] Out of those people, only a handful of innovations prove transformative enough to merit deployment. But this volume isn't waste; it's the price of finding the exceptional. There are various tactics to generate this volume, which I'll explore next.

Phase 3: Validating Ideas and Enabling Experimentation

Before generating solutions through any mechanism—hackathons, workshops, or organic discovery—tools are required to ensure that work is being done on the ideas that matter. Whether at the beginning of ideation or later when evaluating proposals for funding, two complementary tools help teams find innovations that are worth pursuing.

The first tool is *problem reframing*. Take a problem statement, then ask why you need to solve this. Keep asking "why?" to move up abstraction levels or ask how to move down to specifics. Each level opens different solution possibilities. The right problem level passes three simple tests.

1. First, you need Control, meaning you can actually influence the outcome. A solution for the prompt "improve humanity" is not possible, but solving "reduce customer complaints by 30 percent" is doable.

2. Second, you need Multiple solution paths visible. If only one can be determined, the process is too narrow. If there are infinite possibilities, things are probably too broad.

3. Third, you need Measurable impact. If success can't be defined concretely, the right level has not been found.

For example, a logistics team might start with this problem statement: "Use AI to reduce delivery delays."

They reframe it by asking a series of simple "why" and "how" questions to move up and down abstraction levels:

Why? Because delays create customer complaints.

Why? Because customers don't know when drivers will arrive.

That reframing leads to a clearer, more actionable problem: "Provide customers with accurate, real-time delivery ETAs."

This version gives the team control, opens multiple solution paths, and defines measurable impact.

The second tool is Amazon's Working Backward method, specifically their Future Press Release test. Before building anything, write—as if the product had already been launched successfully—a press release. If a compelling announcement that customers would care about can't be written, then there isn't something worth building. The press release forces brutal clarity:

- What does the customer get?

- Why should they care?

- How is it different?

The FPR becomes a north star. If the team ever gets lost in development, they can return to the press release to remember what matters. Amazon takes this a step further with FAQ documents that force teams to answer the hard questions upfront, such as "Why will this fail?" "What dependencies could kill it? and "How much will customers really pay?"

These aren't sequential gates, but filters that can be applied at any stage. Some teams use them as innovation challenge prompts, asking participants to write the press release for their idea. Others apply them during reviews, reframing problems when solutions feel stuck. The key is using them before significant investment has occurred. They're equally valuable at killing bad ideas early as they are at sharpening good ones.

Once an idea survives these filters, teams need somewhere to experiment. Most enterprise sandboxes are useless because they lack data, the one thing innovators need most.

AI innovation is dead on arrival without broad data access. An internal data platform should have standardized APIs for:

- All core business data.

- Synthetic data generation capabilities for safe development.

- A clear data catalog showing what exists and how to access it.

- Automated compliance checks are built into the access layer.

- Usage tracking to identify which datasets are being used and by whom.

Effective innovation infrastructure requires isolated cloud environments provisioned in hours rather than weeks, synthetic data that mirrors production patterns without privacy risks, pre-integrated AI and ML tools, and hard spending limits (e.g. $5,000) that prevent runaway costs. Without these, your sandbox becomes a bureaucratic obstacle rather than an innovation accelerator. When getting a sandbox requires two weeks and five approvals, innovation will happen elsewhere, usually on local machines with downloaded customer data that just creates the security nightmares that sandboxes were meant to prevent.

Once technical infrastructure and validation tools are present, it's time for systematic ways to generate ideas. The principle of volume requires different mechanisms for different operating models and cultures. Innovation challenges or competitions work well for specific problems, with a clear challenge, broad open participation, modest resources for exploration, and systematic filtering. These create energy and focus, though they require clear problem definition upfront.

For ongoing innovation, continuous submission systems work better. Create simple portals for idea submission, make them frictionless, provide rapid feedback, and track everything for patterns to catch ideas that emerge organically but require constant management. Cross-functional workshops bring diverse perspectives together when bridging silos is imperative. Gather people from different departments, focus on specific opportunity areas, use structured facilitation, and produce actionable outputs to generate breakthrough thinking, even if it requires significant time investment.

Also, innovation time allocations, like Google's famous 20 percent, give employees permission to explore. Set aside dedicated time for innovation, provide resources and tools, create sharing mechanisms, and celebrate both successes and failures to build innovation muscles throughout the organization, though this will require cultural commitment. External partnerships expand innovation capacity beyond internal resources. Work with startups for cutting-edge capabilities, partner with universities for research, engage customers in co-creation, and tap into global talent through open innovation. These bring fresh perspectives, but require careful IP management. Choose mechanisms that fit your operating model. Hub-and-spoke organizations excel at challenges that centralized teams can coordinate. Federated organizations do better with continuous submission and local workshops. Venture studios focus on external partnerships and dedicated innovation time.

Don't ignore the compliance element while creating an innovation engine. Making every experiment navigate full compliance will slow things down, but one serious regulatory violation will eliminate decades of innovation ROI. The solution is building compliance into the infrastructure rather than the process. Automated privacy checks should run for any data access, audit trails should be created automatically for all experiments, and pre-approved tool stacks should already be vetted for security and compliance. The development of clear boundaries defines what will require additional review versus what can proceed within existing guardrails.

Phase 4: Governing the Innovation Portfolio

Innovation needs boundaries, not blueprints. Google's OKR system demonstrates how to provide strategic alignment, while preserving team autonomy. Teams know what success looks like, but can choose their own path to get there.

Define innovation zones that make it clear "where" things will play out, meaning which markets, technologies, and customer segments. Articulate how you'll win by leveraging unique organizational advantages. Finally, define success metrics that make success measurable and unambiguous.

Strategic alignment sets direction, but resource allocation determines what actually gets built. Successful innovation requires balancing the portfolio across risk levels. Consider these three innovation categories, based on how far things are being stretched from the business's core:

- **Incremental innovations**: These improve existing products for current customers, making the existing business better.

- **Adjacent innovations**: These are about either new markets *or* new products—but not both simultaneously. For example, *either* selling an existing product to a new customer segment or creating a new product for current customers.

- **Transformational innovations**: These venture into both new products *and* new markets, territories where no existing advantage exists. The risk and reward increase exponentially when moving from incremental to transformational.

Allocate roughly 70 percent of innovation resources to incremental improvements, which are low risk, return modest rewards, and are defensive in that they maintain competitive parity. They seldom do more than pay for themselves, but are necessary to keep pace in the marketplace.

Invest 20 percent in adjacent opportunities that involve either new products or new markets. Such medium-risk, medium-reward bets can provide the best return on innovation investment. They build on existing capabilities, while exploring new opportunities.

Reserve 10 percent for transformational innovations that involve both new products and new markets. These high-risk moonshots have a 95 percent probability of failure, but that 5 percent that succeed can transform a business. Many of these transformational innovations emerge from lessons learned during failed experiments.

Traditional budgeting, with quarterly reviews, creates perverse incentives to overpromise and undercut. Instead, adopt venture capital-style tranche funding, which releases resources based on learning rather than time. Start with seed funding of $10,000 to $50,000 to validate the problem and solution approach. If that succeeds, Series A funding of $100,000 to $500,000 enables building a prototype and testing with customers. Series B funding of $1 million to $5 million supports piloting with real customers in controlled environments. Finally, Series C funding of $5 million or more enables scaling to full deployment. Each tranche requires evidence of learning, not just progress. Measure which assumptions were validated, what external feedback was collected, and what's now known that wasn't before. Not activities completed or money spent. This forces teams to learn quickly rather than perfect plans, controlling risk while maintaining velocity.

None of this works, however, without the right metrics. Most organizations measure innovation incorrectly by counting only successful launches, while ignoring what can be learned from killed projects. These "tailpipe measures" mean that, by the time revenue results, it's three to five years too late to fix innovation problems.

Two metrics matter universally. First, innovation velocity, which is "How fast do you move from idea to validated pilot?" While traditional companies take 18 months to test an idea, digital natives measure this in weeks. Second, what percentage of your revenue comes from products or services launched in the past three years? If it's below 20 percent, innovation is not happening fast enough to stay relevant. 3M famously targets 30 percent from the past four years, a metric that forces honest conversations about whether innovation is actually reaching customers or just filling PowerPoints.

Don't demand positive ROI projections for early-stage innovations. The iPhone would have failed any reasonable ROI calculation. Measure learning speed first, financial returns later.

Even the best funding model and metrics fail if no one owns the transition from innovation to operations. Many pilots die post-validation because the innovation team declares victory and moves on, while the operations team, which wasn't involved and doesn't want the responsibility, is suddenly left holding the bag. The innovation becomes an orphan. Amazon solved this with single-threaded leadership, one person owning an initiative from conception through scaling. No handoffs mean no dropped balls; no divided accountability means no finger-pointing when things go wrong. An "innovation integrator" ensures promising ideas don't die in the gap between innovation and operations.

Phase 5: Sustaining Innovation Through Culture

Most organizations punish failure, which deactivates innovation. Smart organizations reframe failure as learning opportunities rather than career-ending mistakes. This changes the psychology from binary success/failure to degrees of learning.

3M Post-it Notes famously grew from a failed attempt to create a super-strong adhesive. The weak adhesive was useless for its intended purpose but perfect for removable notes. Intuit holds "failure parties" that celebrate bold attempts that didn't work out but which led to captured learning. "At Intuit we celebrate failure," says co-founder Scott Cook, "because every failure teaches something important that can be the seed for the next great idea."[5]

Reframing failure is just the initial cultural shift. The fundamental challenge is that optimization and innovation require opposite organizational muscles, as discussed at the beginning of this chapter. Optimization minimizes variation, reduces risk, and perfects existing processes; innovation embraces variation, accepts risk, and explores new territories.

Companies that excel at both don't try to blend these cultures. They create clear boundaries between the *exploit* (running the business) and the *explore* (building the future). Different metrics are used for each. Operations focuses on efficiency, quality, and hitting targets, while innovation focuses on learning speed and breakthrough potential. Innovation budgets are protected when quarterly pressure hits. There is no raiding the innovation fund to hit earnings targets. Different wins are celebrated: *execution* for operations, *learning*, even from failures, for innovation.

Finally, if innovating isn't a path up the ladder, then the best innovators will leave for companies where it is. Most major tech companies now offer dual-ladder systems where technical experts can reach executive-level status and compensation without managing anyone. Without such paths, innovators face the tough choice of abandoning innovation to advance their careers or leaving for organizations that value their contributions.

Case Study: How Innovation Made DBS the World's Best Bank

In 2009, DBS Bank faced an uncomfortable truth. As Singapore's largest bank, it had spent years optimizing operations; however, customer satisfaction scores remained the worst among the country's banks.[6] Chief data and transformation officer Paul Cobban recalls it was "almost embarrassing to tell people at dinner parties" that he worked there because DBS had such a bad reputation.[6] Long lines snaked through branches. Credit card applications took weeks.

The bank had operational capability and was able to process millions of transactions efficiently, but that wasn't enough to make for good customer experience.

If DBS had fallen into the efficiency trap, consultants would have arrived with process optimization playbooks. They'd benchmark against peer banks, consolidate branch networks by 20–30 percent, and set aggressive cost-to-income ratio targets. The bank would reduce headcount, cut costs by 20 percent, and match peer performance. Efficiency metrics would turn green.

However, DBS did something radically different. It invested heavily in innovation while fixing operations. The aspiration wasn't just to satisfy efficiency metrics, it was to operate like tech giants and compete on innovation, not just cost. This wasn't marketing fluff; it was a systematic transformation.

One of their first successes wasn't adding technology; it was removing complexity. Five-day process improvement events (PIEs) took place, and these workshops discovered significant waste and inefficiency throughout their processes. The PIEs eliminated 250 million customer hours.[6] Credit card replacements, which had taken weeks, now took days as bureaucracy was eliminated.

The bank established a dedicated innovation infrastructure:

- **DBS Asia X (DAX)**: An innovation facility at which employees could collaborate with fintechs in project pods and co-working spaces.[7]

- **Digital platforms mapped to business segments (33 in total)**: Each had a "two-in-a-box" leadership model, led jointly by someone from business and someone from technology.[8]

- **ALAN platform**: Reduced AI deployment time from 18 to under 5 months.[8]

- **Digital Value Capture (DVC)**: A measurement framework that quantified exactly how much digital innovation delivered. The data showed digital customers were dramatically more valuable: they generated double the income, carried significantly higher balances across deposits, loans, and investments (1.5x, 2x, and 3.6x respectively), and cost 57 percent less to acquire than traditional customers.[6]

DBS embedded the behaviors that power innovation directly into its performance management system. The bank dedicated 20 percent of its balanced scorecard to digital transformation, meaning these priorities directly affected how people were evaluated and paid. It also codified cultural expectations around agility, learning, customer obsession, data-driven decisions, and experimentation, making these traits a requirement for everyone.[6] By hardwiring these expectations into how people were evaluated and developed, DBS made innovation a daily operating discipline instead of a side project.

Since 2015, the bank ran over 1,000 experiments across all departments, with an understanding that volume is required to find the breakthroughs.[9]

The results tell a story that efficiency alone never could. DBS achieved return on equity of 18.0 percent, one of the highest among developed market banks globally, while maintaining a cost-income ratio of 40 percent.[10] Digital customers delivered 39 percent ROE—15 percentage points higher than traditional customers—while costing 50 percent less to serve. AI initiatives alone generated S$150 million in additional revenue.[8]

By 2018, Global Finance named DBS "World's Best Bank," and Euromoney awarded it the "World's Best Digital Bank" designation multiple times.[6] These weren't participation trophies; they reflected fundamental transformation from a regional bank struggling with customer satisfaction to a technology company succeeding at cutting-edge banking services.

Notice that DBS didn't abandon operational excellence to chase innovation. And crucially, innovation wasn't a side project run by a distant team, it was built into how every business unit operated, measured, and got paid.

Efficiency made DBS competitive; innovation made it the "World's Best Bank."

Days 1-30: This week, choose your operating model; if in doubt, default to hub-and-spoke. Ring-fence 0.25–0.50 percent of revenue as a protected innovation budget that is reported directly to you or the board, immune from quarterly cuts. Identify three people who've already built something before. These are your first innovation leaders. Look for T-shaped individuals: deep expertise in one domain combined with broad collaborative skills. Post an initial innovation challenge company-wide, like a specific $10 million problem, and open submission for four weeks, with $10K–$50K seed funding available for selected ideas. Set up three collection mechanisms: submission portal, cross-functional workshop, and solicitation of innovations that employees are already building (with or without official approval). Run prune diagnostic on the five most complex processes. Target 100+ ideas by Day 30.

Days 31-60: Require a Future Press Release for any idea seeking funding: what does the customer get, why should they care, how is it different? If a compelling announcement can't be written, don't fund it. Fund the top 10 initiatives with seed funding of $10,000–$50,000 each. Allocate portfolio following 70–20–10 across incremental, adjacent, and transformational. If operational health is weak, focus on prune innovation until the foundation is sound. Define innovation zones: which markets, technologies, and customer segments are in scope. Build infrastructure: provision sandboxes using synthetic data and pre-approved tools, establish a data platform with standardized APIs, and automate compliance into the access layer. Validate five to seven initiatives with clear funding gates by Day 60.

Days 61-90: Release Series A funding only for initiatives with a validated problem and solution approach at $100K–$500K. Establish two permanent dashboard metrics: innovation velocity (idea to validated pilot under 90 days) and revenue percentage from products launched in the past three years (target 20 percent or more). Assign single-threaded owners to each funded innovation, one person from prototype through production. Publicly celebrate first intelligent failure. Conduct a first quarterly portfolio review to fund winners, kill zombies, and celebrate learning.

What Happens Next: Stay vigilant: when your best innovators start leaving, when ideas sit unanswered in submission portals, when pilots multiply but nothing reaches production, these are warning signs that require attention. Continue tranche funding tied to evidence of learning, not time elapsed. Embed innovation into performance scorecards at 20 percent weight. Create dual career tracks via which principal innovators can achieve VP-equivalent status without managing people. Protect innovation budgets when earnings pressure hits; if innovation funding doesn't survive Q4, it was theatrical, not strategic.

[1] Clayton M. Christensen. The Innovator's Dilemma: When New Technologies Cause Great Firms to Fail. Boston: Harvard Business School Press, 1997.

[2] "Innovation Lessons from Pixar: An Interview with Oscar-Winning Director Brad Bird." McKinsey Quarterly. April 1, 2008. https://www.mckinsey.com/capabilities/strategy-and-corporate-finance/our-insights/innovation-lessons-from-pixar-an-interview-with-oscar-winning-director-brad-bird

[3] Gabrielle S. Adams, et al. "People Systematically Overlook Subtractive Changes." Nature 592 (2021): 258–261. https://www.nature.com/articles/s41586-021-03380-y

[4] Alice Piras. "Microsoft Global Hackathon 2025: MVPs Driving Innovation Across Communities." Microsoft Community Hub. August 12, 2025. https://techcommunity.microsoft.com/blog/mvp-blog/microsoft-global-hackathon-2025-mvps-driving-innovation-across-communities/4442513

[5] Ricardo Viana Vargas, Edivandro Conforto, and Tahirou Assane Oumarou. "Making Failure Work." Brightline Initiative. August 15, 2019. https://www.brightline.org/resources/making-failure-work/

[6] David Kiron and Barbara Spindel. "Redefining Performance Management at DBS Bank." MIT Sloan Management Review. March 26, 2019. https://sloanreview.mit.edu/case-study/redefining-performance-management-at-dbs-bank/

[7] "DBS Asia X: fresh innovation at a galaxy not so far away." DBS Innovates. https://www.dbs.com/innovation/dbs-innovates/spaces-dax.html

[8] "DBS: Transforming a Banking Leader into a Technology Leader." McKinsey & Company. 2024. https://www.mckinsey.com/capabilities/tech-and-ai/how-we-help-clients/rewired-in-action/dbs-transforming-a-banking-leader-into-a-technology-leader

[9] David Gledhill. "Reimagining Banking: Why Big Banks Must Operate Like Start-Ups to Succeed." DBS Innovates. June 8, 2017. https://www.dbs.com/innovation/dbs-innovates/reimagining-banking-why-big-banks-must-operate-like-start-ups-to-succeed.html

[10] Chng Sok Hui. "Annual Report 2024: CFO Statement." DBS Bank. 2024. https://www.dbs.com/annualreports/2024/cfo-statement.html

Epilogue

We covered a lot of ground in this book. Your organization will be at a different stage with each strategy. Come back to the ones that need the most work, but don't wait to perfect one before working on the others. Mastering the 8 Strategies together will give you a firm grip on AI transformation that I hope will help you turn your vision into impact.

Go build the future.